Fast Car Physics

CHUCK EDMONDSON

THE JOHNS HOPKINS UNIVERSITY PRESS
Baltimore

Printed in the United States of America on acid-free paper
2 4 6 8 9 7 5 3 1

The Johns Hopkins University Press
2715 North Charles Street
Baltimore, Maryland 21218-4363
www.press.jhu.edu

Library of Congress Cataloging-in-Publication Data

Edmondson, Chuck.
Fast car physics / Chuck Edmondson.
p. cm.
Includes bibliographical references and index.
ISBN-13: 978-0-8018-9822-8 (hardcover : alk. paper)
ISBN-10: 0-8018-9822-6 (hardcover : alk. paper)
ISBN-13: 978-0-8018-9823-5 (pbk. : alk. paper)
ISBN-10: 0-8018-9823-4 (pbk. : alk. paper)
1. Automobiles—Dynamics. 2. Automobiles—Equipment and supplies.
3. Automobiles—Design and construction. 4. Physics. I. Title.
TL243.E36 2011
629.2′31—dc22 2010017554

A catalog record for this book is available from the British Library.

Special discounts are available for bulk purchases of this book. For more information, please contact Special Sales at 410-516-6936 or specialsales@press.jhu.edu.

The Johns Hopkins University Press uses environmentally friendly book materials, including recycled text paper that is composed of at least 30 percent post-consumer waste, whenever possible. All of our book papers are acid-free, and our jackets and covers are printed on paper with recycled content.

Contents

Preface

Before I can pull the retractable ladder down from the ceiling of my garage and retrieve our Christmas decorations from the attic, I have to move a stack of tires. I sit and stare at two full sets of wheels and five sets of tires and wonder, how did this happen? Between the tools and tires, there is no room for the car and it's too cold outside for banging knuckles. The season opener is three months away and I would rather tinker with the car than hang decorations. I would rather plan the next car modification than work on the list of chores that I postponed over the course of last racing season. The Sports Car Club of America (SCCA) calls my addiction "solo racing." Everywhere else, it's called "autocross." When the weather is nice and the opportunity presents itself, this is the way my family spends much of its free time.

I have always been a car nut. As a kid, I drove 500 miles to watch Mark Donahue beat Richard Petty in the first season of the International Race of Champions. As much as I would have loved to follow in Donahue's path, school, work, family, and a modest income limited my racing activities to observing. When my son reached driving age and racked up his first few speeding tickets, I knew he needed a way to get the urge to drive fast out of his system, and autocross fit the bill. It is fast, safe, exciting, and broken into classes to make everyone competitive. It's hard to beat the thrill of racing your daily driver flat out. At first, it was only my son pushing his 1990 Honda CRX Si to the limit. When I finally reached the point where I could afford a Nissan 350Z, I could no longer pass up the temptation. I had a blast. Somewhere around the 220,000-mile point, the CRX gave up the ghost, and my son and I co-drove the Nissan for a season. Soon, my daughter joined the act. Every successful team needs a patient and generous sponsor to provide support and cash for new parts. Ours can be found smiling trackside and in the pits. She also doubles as my wife.

When I started autocrossing, I ran in the B stock class and, to be honest, wasn't exactly winning my class. To add insult to injury, the SCCA uses a handicap system called the PAX index, and it allows comparison of the various classes. It is based on the relative speeds at national events in each class. At the end of the day, all 200 or so cars that race in my region are ranked in order. Although I did not expect to win, I didn't expect to be thrashed. I don't take thrashing well, and I don't like to quit. I needed to fix my driving and to learn how to set up my car. This doesn't happen overnight. It takes experience and understanding. It turns out that a little money is useful as well. A year and a half later, I managed a finish of 29th place out of 122. I won my class that

The Team Paco's Parts House 350Z out on the autocross track. Photograph by Clyde Caplan and Alex Teitelbaum.

day and just edged out a BMW M3 for the season class championship. It was a small class in a small club, but it was a sweet victory nonetheless. My car was co-driven by one of the top drivers in the region and, by raw times, only three cars were faster. He was 0.4 seconds behind the *f*astest *t*ime of the *d*ay (FTD). Progress is possible.

Automotive advice fills the Internet and the library, much of it qualitative, some of it wrong. The quantitative stuff written by engineers is often intended for other purposes. As a physics professor and a former Navy nuclear engineer, I needed something more. I set about the process of trying to understand racing from the perspective of physics. My new hobby soon became a one-credit seminar, and then two, and finally a three-credit undergraduate course. This book is a compilation of much of what I learned along the way. It explains racing in terms of basic physics. Most of the physics is directed at the high school and college freshmen level. If you hate all forms of mathematics, this is not your text. If you can use a little algebra and trigonometry, you'll be just fine. If you can run a spreadsheet program, you can even apply the most complex ideas to any car of your choosing. I'll highlight the ideas and conclusions that you can apply to your own car. Hopefully, by the time we are done you'll understand what happened to my garage.

Acknowledgments

Many people have made this text and my time behind the wheel possible. I owe them all a sincere debt of gratitude. First and foremost, I would like to thank the owner and my co-drivers at Team Paco's Parts House. The owner is the love of my life and my wife JoAnne. My co-drivers are my son Tristan and daughter Brittany. With the three of them at my side, every day of racing and driving is a joy. My folks, George and Mary Edmondson, put up with many years of racing magazines and car repairs. My Dad taught me how to turn a wrench, a necessary skill for any racer that isn't independently wealthy. I owe my family a great deal.

Clyde Caplan and Alex Teitelbaum generously supplied the photos in this text. They are photographers and de facto historians of Team WTF?!

The men and women of the Washington DC Region of the Sports Car Club of America are a wonderful group. Their solo program (autocross) is one of the top in the nation. I have also had the privilege of serving on the organizing committee of the DC Region Time Trials and Performance Driving Experience (PDX). I have learned from these two programs how to engage in safe and fun high-performance driving. Friendly, open, and ready to help, they are simply a great group. The region club racers, flaggers, stewards, and other officials volunteer to make the PDX program possible. I am also a member of Autocrossers Incorporated, one of the premier autocross clubs in the country. As a member, I have had the privilege of working with, watching, and learning from some of the best drivers that I have known. I am proud to call them my friends.

From the moment that I landed in the Physics Department of the United States Naval Academy, I have felt at home. They have enthusiastically supported the development of my course on the physics of motorsports and this text. Thank you to all of the faculty and technical staff of our department. I owe a special thank you to a few people. Professors John Fontanella and Mary Wintersgill have supported my development as a faculty member at every turn. John has suffered through countless discussions on physics and racing. Professor Don Treacy hired me as a civilian professor and provided many helpful insights upon reviewing the draft of this text.

There are so many others that I cannot mention them all by name, but I would like to conclude by thanking Trevor Lipscombe, editor-in-chief at the Johns Hopkins University Press. He has shown me endless patience.

Fast Car Physics

Chapter 1

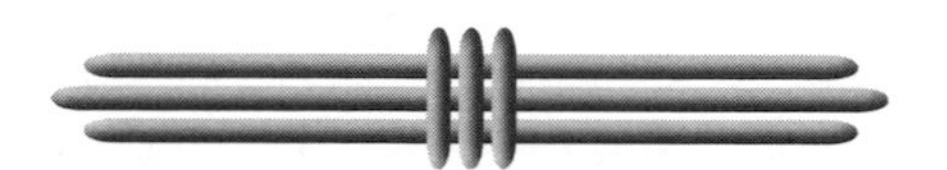

Torque or Horsepower? Finding the Shift Points

I was watching an international broadcast of A1 Grand Prix involving 550 horsepower open-wheeled racers. Each car represented a different country. The British announcing team tossed around the names of European drivers and gave the show an air of intelligence. Their apparent knowledge and self-confidence were impressive. An in-cockpit camera revealed a bar of lights that grew in length as the engine revved up. The colors of the bar transitioned from green to yellow and eventually to red, giving the driver a visual clue about when to shift gears. The announcer chimed in by saying, "I've driven one of these cars without the lights and it's very difficult to know when to shift. With the lights, you end up shifting at the peak of torque and horsepower!" This brought a smile to my face. He was, as we will soon see, wrong. He may have driven the car, but he failed to understand the technology behind the car.

The great drivers of today, the Michael Schumachers, Jeff Gordons, and Andrettis, are part of the car development team. They understand the science and technology, and they provide the kind of feedback that leads to consistent improvement. If you do your homework, imposters are easy to spot. While A1 Grand Prix involves road racing on some of the most complex tracks in the world, it is easier to understand our announcer's gaffe by considering the straight-line motion of drag racing. We will explore the relations between the twisting force, called torque, and acceleration, and we'll develop a strategy for selecting shift points when racing. We will reveal how the rate of doing work, expressed as horsepower, is related to the top speed attainable by our car.

On open-track night at the local drag strip you'll be tempted to take a turn behind the wheel. Most of us resist and keep our car in the parking lot, but many give in. Everything from the finest American muscle car to nitrous-fed tuners and Mom's minivan seem to be lined up for the chance to race the clock. The minivans, sedans, and some of the performance cars have automatic transmissions. If you are a serious enthusiast, you probably have a manual transmission, which leaves the decision of when to shift gears in your hands. If you ask the amateur racers about shifting strategy, you'll get a wide variety of answers. By amateurs, of course, I mean everyone else at open-track night. I'm not talking about the folks that make their living at it. They face problems that probably do not apply to our cars. Often they are limited by having too much power for the track conditions, a nightmare that few of us dare to dream. The rest of us with moderate torque and power are free to pick any shift-point rpm that lies between stalling out and blowing up.

There are three common shift strategies given by veterans of open-track night: (1) shift at or just beyond the peak in the torque curve, (2) shift at or just beyond the peak in the horsepower curve, or (3) shift at the redline. The terms *torque curve* and *horsepower curve* mean the quantities plotted as a function of engine rpm (revolutions per minute). The third strategy is the easiest of the three to follow. You probably have a tachometer, commonly called a tach, and there is a red mark on it someplace, typically at around 6000 or 7000 rpm. When you reach the red mark, shift gears. It's that simple, if you remember to watch the tach. Those addicted to the track frequently use a tach with a bright red shift light or series of lights to catch their attention. This is particularly true of those who have blown up an engine or two by going beyond the redline. I remember seeing a Porsche 911 engine that had suffered from a missed downshift. The driver had planned to shift from fifth to fourth and instead hit second gear. Some analog tachometers have a needle, called a telltale, that stops at the highest rpm that is reached. The telltale on the Porsche tach showed a maximum engine speed about 2000 rpm beyond the redline. At this engine speed, valves and pistons collide, leaving behind large quantities of scrap metal.

The first two shifting strategies assume some higher degree of specialized knowledge. You must have found a book, magazine, or Web site that shows the torque and horsepower curves for your car. At least, you must find the peak values quoted somewhere, for example, 200 hp at 4000 rpm. If you are fortunate, you have access to a dynamometer that will provide customized curves for your car in its current state of modification and wear. We'll discuss the dyno a little later. Nevertheless, even if you have the "dyno" data, which shifting strategy is correct? The choices yield drastically different results. To explore the physics answer to this question, we're going to concentrate on the 2004 Subaru WRX STi. Why? The STi is a rocket and there is a ton of easy-to-find data. Subaru tells us that the STi has an engine torque peak at 4000 rpm, an engine horsepower peak at 6000 rpm, and a redline at 7000 rpm. A chassis dyno measures forces at the wheels and will tell us how much of the engine torque and horsepower actually make it through the drivetrain of the car and to the ground. Because drivetrain losses increase with rpm, you will find that the horsepower peak slips from 6000 to 5500 rpm on a chassis dyno. The dyno numbers that we are going to use were extracted from a graph found in an August 2003 *Sport Compact Car* article. It is truly amazing how much of this type of information is available to us. You can follow this approach for any car of your choosing and analyze its performance.

Our possible track night strategies, when applied to the STi, leave us with a band of advice that ranges from 4000 to 7000 rpm! To be honest, the engine is really only useful from about 2000 to 7000 rpm. The uncertainty of when to shift covers 60% of the useful rpm range. Clearly, someone's strategy is wrong.

1.1 ACCELERATION AND NEWTON'S SECOND LAW

In introductory physics, we describe the motion of objects with three vector quantities: position, velocity, and acceleration. Position is the location of the object relative to some coordinate origin. Because it's a vector, it has a magnitude in feet or meters or miles and a direction. In straight-line motion, the direction is given by the positive or negative direction, and in drag racing, the finish line is the positive direction. Velocity is the rate of change in position, in other words, how fast is it going and in what direction? The acceleration is the rate of change in velocity. We call speeding up positive acceleration and slowing down negative acceleration. The fastest trip down the quarter mile will be achieved when the car maintains the greatest possible acceleration, *a*, throughout the trip. Forces cause acceleration. The relationship between force and acceleration leads us to Newton's second law.

Newton's Second Law

Newton's second law relates acceleration, a, to the net applied force, F_{net}, that acts on the car. How much acceleration? This depends on the car's inertia or resistance to acceleration. This inertia is referred to as the mass, m.

$$F_{net} = ma \text{ or } a = \frac{F_{net}}{m}, \tag{1.1}$$

where F_{net} is the total force in pounds (lbs), m is the mass in slugs, and a is the acceleration in feet per second squared (ft/s^2). Again, since drag racing is a one-dimensional problem, the vector nature is contained in the sign of the force or acceleration. Please forgive the use of British units, but most of the data available for cars in the United States are still in nonmetric quantities like ft-lbs and horsepower. We'll review the units as we go. Pounds are easy to understand because everyone's bathroom scale reads weight (which is loosely the force of gravity) in lbs. You may recall that the metric system measures force in Newtons, N, and mass in kilograms, kg. Mass in slugs is obtained by dividing an object's weight in lbs by the free-fall acceleration of gravity, which is 32.0 ft/s^2. F_{net} in Newton's second law is the vector sum of all external forces that act on the car. For the moment, let us neglect the aerodynamic drag and other forces that tend to slow the car and focus only on the force that makes the car speed up. That force is friction, and it acts at the point where the drive wheels contact the road.

Newton's second law can also be used to consider rotation. The analog of a force that causes a rotation is called a torque, τ. A torque requires that a force be applied at some distance from the axis of rotation called a lever arm. The magnitude of the torque is equal to the product of the length of the lever arm, in feet, times the component of the force that is perpendicular to the lever arm, in pounds. The torque on the car wheel due to friction is equal to the radius of the tire times the frictional force in units of foot-pounds. The twisting force, or torque, causes angular acceleration, α, in radians per second squared. How much angular acceleration exists? That depends on the rotational inertia, I. In physics, the rotational inertia is called the "moment of inertia." It depends on how much mass is rotating and how far that mass is from the axis of rotation. We'll come back to the details later. Newton's second law for rotation is written as

$$\tau_{net} = I\alpha.$$

It is not a coincidence that the language that describes rotation is the same as the language of linear motion.

To analyze the shift points, we must be able to express the acceleration as a function of engine rpm and the speed of the car. We will start at the tire and work our way back to the engine.

Figure 1.1 shows the frictional force that the road exerts on the tire, F_{RT}, as the tire rotates clockwise. This force arises in response to the twisting force, called a torque, which is applied to the tire and wheel by the axle. When the tire tries to spin, the friction builds and opposes the motion. It exerts a torque

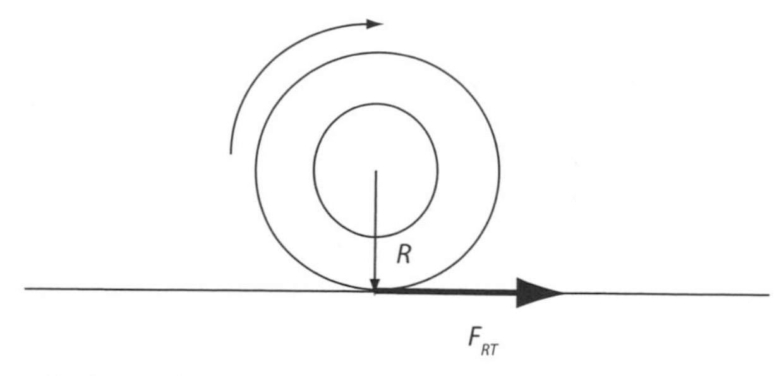

Figure 1.1 The force of the road on the tire, F_{RT}, is the force that accelerates the car.

about the center of the wheel with a lever arm equal to the radius of the tire, R. Strictly speaking, the lever arm is the distance from the axis of rotation to the point of application of the force. The tire is not perfectly round, so the lever arm is less than the average tire radius. As a first approximation, it is good enough. The torque acting on the wheel, τ_W, due to friction acting on the tire is

$$\tau_W = RF_{RT} . \tag{1.2}$$

The symbol τ_A represents the torque applied by the axle. Applying Newton's second law for rotation to a drive wheel and tire,

$$\tau_{net} = \tau_A - \tau_W = I\alpha . \tag{1.3}$$

The torque applied by the axle minus the torque applied by the ground, τ_W, is equal to the rotational inertia of the tire and wheel, I, times their angular acceleration, α. For the time being, let's assume that rotational inertia is small.

The torque at the wheel, τ_W, is significantly different from the engine torque, τ_E. There are gears between the engine and wheel. These gears, some of which are located in the transmission and some in the differential, have two functions. First, they slow the rotation rate so that the tires don't have to spin at 7000 rpm. Second, they amplify the torque to make the system more efficient at accelerating the car. If you don't believe this, try starting from a dead stop in third gear. We do this by slipping the clutch and keeping the engine rpm closer to its torque peak. At best, the acceleration will be poor.

The overall gear ratio, G, is the product of the transmission gear ratio, G_T, times the differential gear ratio (sometimes called the final drive ratio), G_D:

$$G = G_D G_T . \tag{1.4}$$

Keep in mind that G_T changes every time you shift gears, with first gear having the largest G_T. The torque at the wheel is the product of the engine torque and the gear ratio:

$$\tau_W = G\tau_E . \tag{1.5}$$

Combining equations (1.1)–(1.5), we get

$$a = \frac{\tau_E G}{mR} . \tag{1.6}$$

We can learn a couple of lessons from equation (1.6). You can maximize acceleration if you increase engine torque. It's not surprising that the most common performance modifications are focused on increasing torque. However, it is neither the easiest nor least expensive. The second lesson is to raise the overall gear ratio. It is common practice in racing to swap the differential gears to match the wheel torque to the track type or conditions. Many track cars are designed with the ability to swap the transmission gear cluster. Third, the cheapest way to gain acceleration is to lower the car's mass. Take out the backseat, the spare tire, the carpet, the sound-deadening materials, the passenger's seat, the CD/radio head unit, amplifier and speakers, the air-conditioning, and the headliner, and you have made a serious dent in the car's mass. Don't forget the gas tank. Gasoline is approximately 6.0 lbs per gallon. If you race with only a quarter of a tank, you may save 60 to 70 lbs. Of course, don't go too low or the fuel pump may lose suction as the gas sloshes around the tank. In theory, weight reduction is the reason that you see so much carbon fiber on tuner cars. Fourth, you can decrease the tire radius. A small-diameter tire seems to go against the current trend in show cars, but the evidence is clear. A 10% reduction in tire radius yields an 11% increase in acceleration. That which looks good on a show car does not mean better performance. We'll come back to these points. For the moment, let's assume that mass, m, and radius of tire, R, are fixed. Keep in mind that G changes with each shift of the gears and the engine torque τ_E is a function of engine rpm.

1.2 VELOCITY, SPEED, ROTATION, AND ENGINE RPM

Now that we have a way to express the acceleration as a function of engine performance, we need to address the question of "when" to shift. For ease of calculation and understanding, most quantities in introductory physics are plotted as a function of time. When you are actually sitting behind the wheel, it is more practical to gage your shift points based on the engine rpm or based on the speedometer. We will also want to be able to express speed as a function of engine rpm for each transmission gear. By doing it this way, we will be able to plot the acceleration as a function of speed for each gear. The gear that gets us the greatest acceleration will be the gear we want to use. The speed of the car, V, is equal to the product of the tire radius, R, times the angular speed, ω, of the tire:

$$V = R\omega. \tag{1.7}$$

We can convert the angular speed, measured in radians per second, into wheel revolutions per minute, rpm_W. It is common practice to refer to angular speed as simply "rpm." We work hard in general physics class to avoid such confusion. One of the biggest challenges when applying physics class work to other disciplines is the process of translating the terminology of the new discipline into the vocabulary of physics. We'll use angular speed and rpm interchangeably. We need to convert the units:

$$\omega\left(\frac{\text{rad}}{\text{s}}\right) \times \frac{1\ \text{revolution}}{2\pi\ \text{radians}} \times \frac{60\text{s}}{1\ \text{min}} = \left[\frac{30}{\pi}\right]\omega\left(\frac{\text{rad}}{\text{s}}\right) = \omega(\text{rpm}_W). \tag{1.8}$$

Solving equation (1.8) for ω and substituting into equation (1.7), we get

$$V = \frac{\pi \times R \times \omega(\text{rpm}_W)}{30} \approx 0.105 \times R \times \omega(\text{rpm}_W). \tag{1.9}$$

If R is given in feet, we have speed in feet per second, which doesn't exactly match the speedometer units:

$$\begin{aligned} V(\text{mph}) &= 0.105 \times R \times \omega(\text{rpm}_W) \times \frac{1\ \text{mile}}{5280\ \text{ft}} \times \frac{3600\ \text{s}}{1\ \text{hour}} \\ &\approx 0.071 \times R(\text{ft}) \times \omega(\text{rpm}_W). \end{aligned} \tag{1.10}$$

The wheel rpm is equal to the engine rpm as read on the tachometer, divided by the overall gear ratio, G:

$$V(\text{mph}) = \frac{0.071 \times R(\text{ft}) \times \omega(\text{rpm}_E)}{G}. \tag{1.11}$$

When we have a particular car and dyno output in mind, we can use equations (1.7) and (1.11) to plot acceleration as a function of speed. Once again, we can learn from considering the form of the equation. Equation (1.11) tells us that decreasing the tire radius will reduce the car's maximum speed for each gear. You'll need to shift sooner if you pick smaller tires. Design decisions in complex equipment often involve making such trade-offs. Which is more important for the function of the car, more acceleration and more frequent shifting or less acceleration and fewer shifts?

1.3 THE CAR, HORSEPOWER, AND TORQUE

We can clarify the complexity of the last section by applying the principles to a specific example. The Subaru WRX STi produces a peak engine horsepower of 300 hp at 6000 rpm and 300 ft-lbs of engine torque at 4000 rpm. You can find these facts in almost any current car data source. A few investigators have taken the car to a dynamometer. A "chassis dyno" measures directly the at-the-wheel forces. By comparing the wheel rotation with the engine rotation, the dyno calculates the overall gear ratio, G, and estimates the engine torque

This quick, nimble, and well-balanced Subaru WRX STi was designed for off-road rally racing and excels in everything from autocross to track days. Photograph by Clyde Caplan and Alex Teitelbaum.

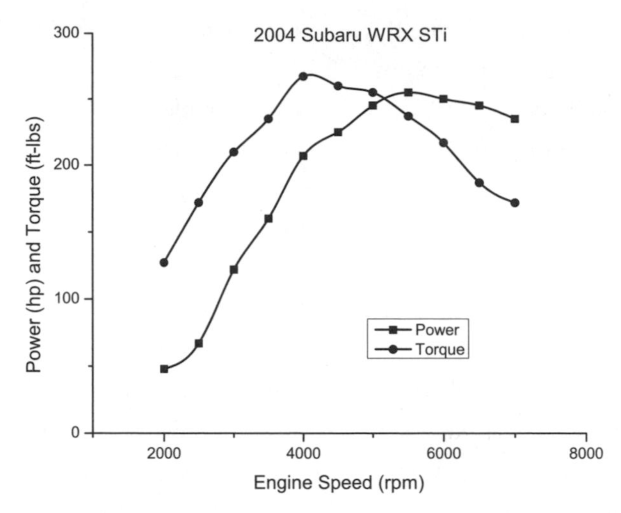

Figure 1.2 Chassis dynamometer horsepower and torque, measured at the wheel, as a function of engine rpm. Data from *Sport Compact Car*, August 2003.

and horsepower as a function of engine rpm. Figure 1.2 shows such a data set found in *Sport Compact Car* in an article written by Josh Jacquot.

Armed with a ruler and calculator, I estimated the values in the published graph. I plugged the values into a spreadsheet program and had it plot a smooth curve. Notice that peak torque and horsepower are well below the advertised 300. However, the manufacturer did not lie. The chassis dyno does not account for internal friction and the rotational inertia of all of the car parts that are forced to spin when you step on the gas. Manufacturers use an engine dyno that measures torque and horsepower at the crankshaft. Subtract the 15%–25% drivetrain losses and you have the useable torque and horsepower measured by the chassis dyno. The raw data are given in table 1.1 if you want to try a few calculations on your own.

The gear ratios found in table 1.2 were obtained from an article in the *Road & Track Road Test Annual for 2004* in an article by Sam Mitani. The article lists the ratios for all six gears, the final drive ratio, and the resulting overall ratios. Since the overall ratios given in the article were not exactly equal to the product of the other two ratios (see eq. [1.4]), we calculated our own overall ratios. Only Subaru knows for sure which of us is right.

TABLE 1.1
2004 Subaru WRX STi Dynamometer Horsepower and Torque

Engine rpm	Wheel Horsepower	Wheel Torque (ft-lbs)
2000	48	127
2500	67	172
3000	122	210
3500	160	235
4000	207	267
4500	225	260
5000	245	255
5500	255	237
6000	250	217
6500	245	187
7000	235	172

Source: Data from *Sport Compact Car Magazine,* August 2003.
Note: Horsepower and torque measured at the wheel, as a function of engine rpm.

TABLE 1.2
2004 Subaru WRX STi Gear Ratios

Gear	Gear Ratio	Final Ratio	Overall Ratio
1	3.64	3.9	14.20
2	2.38	3.9	9.28
3	1.76	3.9	6.86
4	1.35	3.9	5.27
5	0.97	3.9	3.78
6	0.72	3.9	2.81

Source: Gear ratios obtained from the *Road & Track Road Test Annual for 2004.*

The curb weight is 3270 lbs, so the mass is

$$m = \frac{3270 \text{ lbs}}{32 \text{ ft/s}^2} = 102.2 \text{ slugs}. \tag{1.12}$$

1.4 TIRE MARKINGS

We also need the tire radius, R, in feet. Since I don't have an STi handy to measure, I looked up the tire size (thank you *Road & Track*) to calculate R. The size was given as 225/45ZR-17. This is a bizarre mix of units and informa-

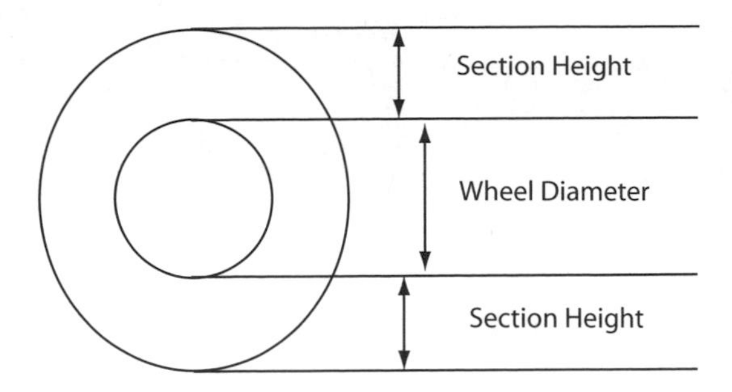

Figure 1.3 Definitions for wheel and tire dimensions.

tion. The 225 is the unloaded, overall tire width in millimeters. Some tires have a sidewall rub strip to protect the tire from curbs. Typically, the rub strip is not part of the 225 mm. Next, 45 is the aspect ratio, meaning that the height of the tire from the rim, called the section height, is 45% of the width. The Z means that it is rated for speeds greater than 149 mph. The R means that it is of radial construction type, and the 17 is the wheel diameter in inches. Figure 1.3 shows the relevant dimensions required to find the radius.

From figure 1.3, it is clear that the tire radius, R, can be found from

$$R = ((2 \times \text{Section Height}) + \text{Wheel Diameter})/2$$

and

$$R = ((2 \times 4.0) + 17) \,/\, 2 = 12.5 \text{ inches or}$$
$$R = 1.04 \text{ ft.}$$

1.5 CALCULATIONS

We are ready to make our first example acceleration and speed calculation. At 2000 rpm in first gear, and using equation (1.6),

$$a = \frac{\tau_E G}{mR} = \frac{127 \text{ ft} - \text{lbs} \times 14.20}{102.2 \text{ slugs} \times 1.04 \text{ ft}} = 17.0 \text{ ft/s}^2$$

$$a = \frac{17.0 \text{ ft/s}^2}{32.0 \text{ ft/s}^2} = 0.53 \text{ g's.}$$

Assuming that the tires don't slip, the car moves off at a little over half of a g of acceleration. It is common practice to express acceleration as a fraction of the

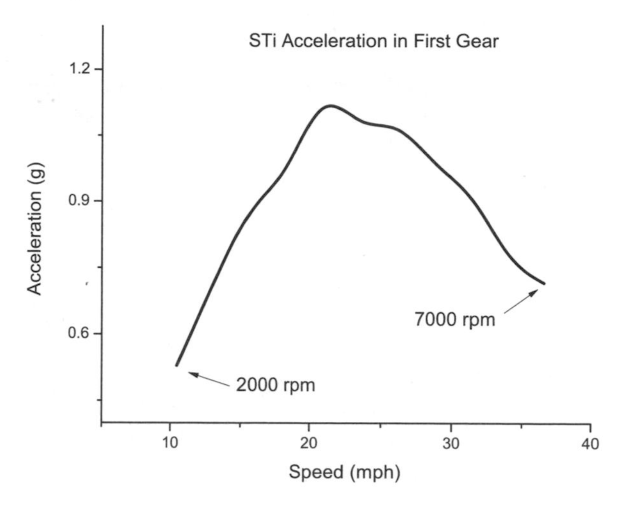

Figure 1.4 First-gear acceleration as a function of speed.

free-fall acceleration due to gravity, which is 32.0 ft/s^2. It gives us a feeling as to whether the acceleration is substantial or not. A typical street tire is limited to about 1 g before slipping. A high-end drag racing tire moving over a patch of asphalt coated with a fresh, hot, black burnout stripe can generate about 4.5 g's. The bigger the g's, the more rapidly the speed increases. Since torque is a function of engine rpm, this acceleration will quickly change.

How fast is the car moving at 2000 rpm in first gear? From equation (1.11),

$$V(\text{mph}) = \frac{0.071 \times R \times \omega(\text{rpm}_E)}{G} = \frac{0.071 \times 1.04 \times 2000}{14.2} = 10.4 \text{ mph}.$$

We'll discuss what happens below 10.4 mph later. For now, we'll move on to 2500 rpm, find a and V, and add this point to our new plot of acceleration as a function of speed. We continue this process until 7000 rpm is completed. The result is plotted as a smooth line for first gear in figure 1.4.

1.6 FIRST GEAR, IT'S ALL RIGHT. SECOND GEAR . . .

Over the range of 2000–7000 rpm, the speed goes from roughly 10 to 37 mph and acceleration goes from 0.53 to 0.72 g's with a peak of 1.11 g's at 4000 rpm. As we expect, this corresponds to the peak in the torque curve. Why not shift

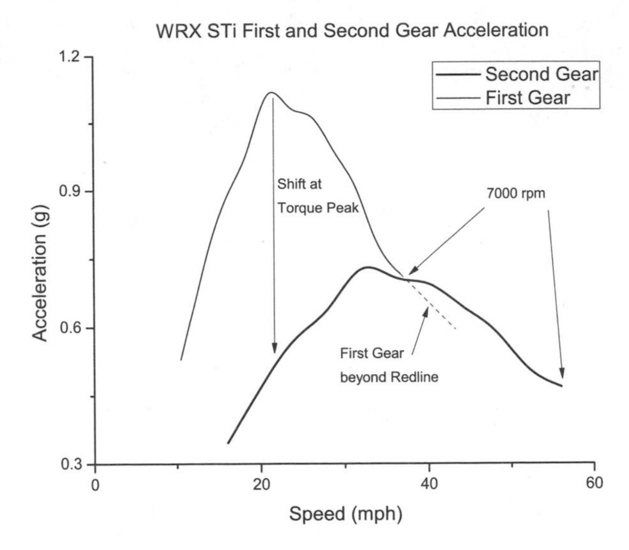

Figure 1.5 Acceleration as a function of speed for first and second gear. The bold vertical arrow shows the drop in acceleration from 1.11 to 0.5 g's experienced by shifting from first to second gear at the peak of the torque curve in first gear. The curves meet at 37 mph where the engine reaches 7000 rpm in first gear. The dashed line is a projection of the acceleration in first gear if engine rpm exceeds the 7000 rpm redline.

here? The answer comes when you add the acceleration curve for second gear to the picture as shown in figure 1.5.

The lower gear ratio in second has reduced the entire acceleration curve but has increased the range of speed. Shifting at the torque peak in first gear would drop your acceleration from 1.11 g's in first to 0.5 g's in second. This would be silly since the entire first gear curve is greater than the second gear curve for any speed. The curves meet at 37 mph where the engine reaches 7000 rpm in first gear. To maximize acceleration, we stay in first gear until redline where the two curves meet.

There is always a temptation to rev beyond the redline, especially in a powerful car with a free-breathing engine. This is bad for two reasons. The redline is based on the mechanical limits of the engine. Different designs have different limiting components, but the point is the same, mechanical breakdown. We find the second reason in figure 1.5. As the engine approaches redline, the torque is falling off. The dashed line is a projection of the acceleration in first gear if engine rpm exceeds the 7000 rpm redline. For the STi, staying in first gear beyond redline will result in a lower acceleration than is available in second gear. There is nothing sacred about 7000 rpm; it just happens to be the limit for the STi. The Honda S2000 currently revs to 8500 rpm, and early editions revved to 9000 rpm. Formula 1 Grand Prix cars rev to 18,000 rpm! The setting of the redline is a function of the mechanical strength, the ability to deliver fuel, air, and oil, and the shape of the torque curve. Figure 1.6 adds the remaining four gears.

Here is where things get interesting. In the shift from second to third, the curves cross at approximately 6470 rpm in second gear, which corresponds to 52 mph. Staying in second gear beyond 6470 rpm or 52 mph will result in an acceleration that is less than can be achieved in third gear. Table 1.3 summarizes the approximate overlap rpm for all five gear shifts.

The results are intriguing and do not under any circumstances match the advice of veterans of open-track night. The redline is reached only in first gear. For every other gear the shift should occur at 400–700 rpm below the redline. If your tach has a shift light and you can only set the light to flash at one rpm,

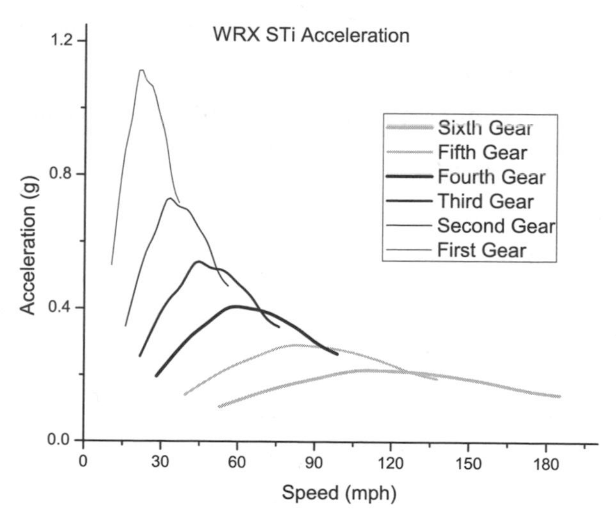

Figure 1.6 Acceleration as a function of speed for all six gears in the Subaru WRX STi.

TABLE 1.3
Shifting Speeds and Engine rpm Required to Optimize Acceleration for the STi

Gear	Shift Speed (mph)	Shift rpm
First	37	7000
Second	52	6470
Third	68	6320
Fourth	92	6570
Fifth	126	6450

I would pick 6500 rpm. If it were fully programmable, I would follow the guidance of table 1.3. Note that all of these shifts are well above the 5500 rpm horsepower peak and 4000 rpm torque peak. There is, in fact, no simple way to determine when to shift. We need knowledge of the gear ratios, tire sizes, and the dyno torque curve to optimize the shift points. Not every car has an overlap in acceleration as we have in the STi. In chapter 2 (section 2.10), we will take a quick look at the Nissan 350Z, which has a flat torque curve and no overlap between the gears.

1.7 SUMMARY

We started this chapter with two goals. The first was to gain an understanding of the importance of torque and horsepower curves. The second was to establish a method for determining shift points for optimum acceleration.

Lessons learned:

- For the stock STi:
  —The shift points as measured in rpm are
  ... beyond the torque peak;
  ... beyond the horsepower peak;
  ... less than the redline rpm.
  —The optimum shift point changes with each gear.
- A falloff in high-rpm torque will exaggerate this effect. Conversely, a little more high-end torque would push the shift point closer to the redline for the STi.

- A higher maximum operating rpm (redline) by itself will not improve the acceleration of the stock STi.
- Smaller diameter wheels produce a larger acceleration and require earlier shift points at lower speeds.
- The grip of stock tires is challenged in first gear. For the remainder of the gears the acceleration is less than 1 g.
- More low-end torque (at low rpm) for the STi would improve acceleration and still not exceed the capacity of the tires. This is where large-displacement engines have an advantage.
- We still have more to learn about horsepower.

Chapter 2

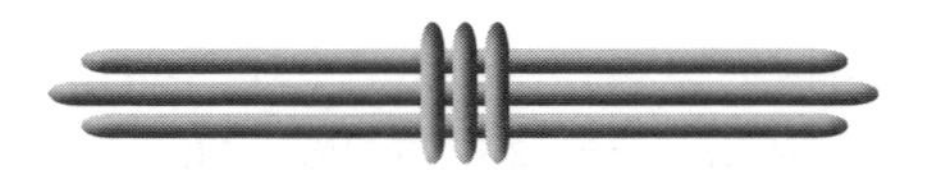

Horsepower, 0 to 60 mph, and the Quarter Mile

2.1 HORSEPOWER

We made it all the way through the discussion in the first chapter without really defining horsepower. Power, P, is the rate of doing work, W/t. Work, W, in physics is a force, F, acting along a distance, d. Combining these definitions,

$$P = \frac{W}{t} = \frac{Fd}{t} = Fv.$$

This assumes that the force, F, and the velocity, v, are collinear (we'll worry about vector products later). It is most easily calculated in drag racing as the product of the force in pounds with the velocity in feet per second. It therefore has the units of ft-lbs/s. A more convenient unit is horsepower. One horsepower is equal to 550 ft-lbs/s. It is common practice to refer to the quantity power as horsepower. The force that is doing this work is the force that is accelerating the car, that is, friction with the road surface, F_{RT}:

$$\text{Horsepower} = (F_{RT}V)/550. \tag{2.1}$$

From Newton's second law $F_{RT} = ma$, where m is the mass of the car in slugs and a is the acceleration in ft/s^2. For the STi, the mass is 102.2 slugs and the acceleration can be obtained from figure 1.6. With each shift of the gears, the velocity increases and the force available to push the car decreases. It goes from a peak of about 3650 lbs in first gear to a minimum of 640 lbs in sixth gear at 145 mph at 5500 rpm. The STi is electronically limited to 145 mph, but what if it weren't? There is still 1500 rpm to reach the 7000 rpm redline, in theory, 185 mph. Unfortunately, it will never get there. Horsepower determines the car's ability to overcome the resistive forces that act on the car. The major factor is the aerodynamic drag. When the rate at which the tires do work is equal to the rate at which drag does work, the car is in equilibrium. At this point, the drag force $f_{drag} = F_{RT}$. So, in a sense the horsepower and the drag force determine the maximum speed of the car. A higher gear won't help raise the top speed because, as we have seen, higher gears actually lower the force available at the tire-road interface.

Horsepower cannot be decoupled from torque. They are closely related. From equation (1.2)

$$F_{RT} = \frac{\tau_W}{R}$$

and from equation (1.7)

$$V = R\omega,$$

where τ_W is the torque of the road acting on the wheel. R is the radius of the tire, V is the speed of the car, and ω is the angular velocity in radians per second, which can also be expressed in revolutions per minute. Plugging equations (1.2) and (1.7) into equation (2.1),

$$\text{Horsepower} = \tau_W\omega/550. \qquad (2.2)$$

The 550 is a unit conversion factor. Hence, horsepower is really the product of the torque, rpm, and a couple of constants to get the units correct. They are closely related, and it is easy to see how the enthusiast becomes distracted by the wrong quantity. Torque is directly related to acceleration, and horsepower is related to top speed.

It is worthwhile to make sure we follow the horsepower units. In introductory physics we have

$$\text{Power (J/s)} = \tau(\text{N} \cdot \text{m}) \times \omega(\text{rad/s})$$

or, in British units,

$$\text{Power(ft} - \text{lbs/s)} = \tau(\text{ft} - \text{lbs}) \times \omega(\text{rad/s}).$$

Conversion factors include

$$1 \text{ horsepower} = 746 \text{ watts} = 550 \text{ ft} - \text{lbs/s}$$

$$\omega\left(\frac{\text{rad}}{\text{s}}\right) \times \frac{1 \text{ rev}}{2\pi \text{ rad}} \times \frac{60\text{s}}{1 \text{ min}} = \omega(\text{rpm}).$$

Using these definitions and conversions, we can relate horsepower, torque in ft-lbs, and ω in rpm:

$$P(\text{horsepower}) = \tau(\text{ft} - \text{lbs})\ \omega\ (\text{rpm})\left(\frac{2\pi}{60}\right)\left(\frac{1}{550}\right) \tag{2.3}$$

$$P(\text{horsepower}) = \frac{\tau(\text{ft} - \text{lbs})\ \omega\ (\text{rpm})}{5252}.$$

When engine rpm is 5252, the torque should equal horsepower. As a quick check for dyno curves from an unknown source, torque and horsepower must cross at 5252 rpm. If they do not, something is wrong.

With these relationships in hand and a few simplified models, it is possible to visualize the relationship between torque and horsepower. For example, if the torque is constant and we apply equation (2.3), we get a constantly increasing power as shown in figure 2.1. If the torque increases linearly with rpm, then the power increases parabolically with rpm as shown in figure 2.2. Note that the entire horsepower figure is concave upward.

At low rpm horsepower increases even though torque is decreasing as shown in figure 2.3. Note that the entire horsepower curve is concave downward. Horsepower peaks and the two curves cross at 5252 rpm. Figure 2.4 portrays a peak in torque represented by an inverted parabola. The resulting horsepower curve is concave up with increasing torque and concave down with decreasing torque. The horsepower peaks later than the torque. The features are similar to those in most real-world curves. In all four cases, the horsepower and torque cross at 5252 rpm.

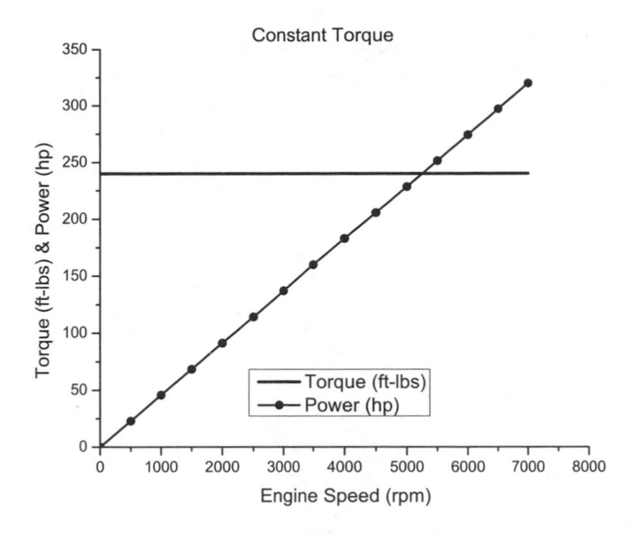

Figure 2.1 Constant-torque results in power that increases linearly with rpm.

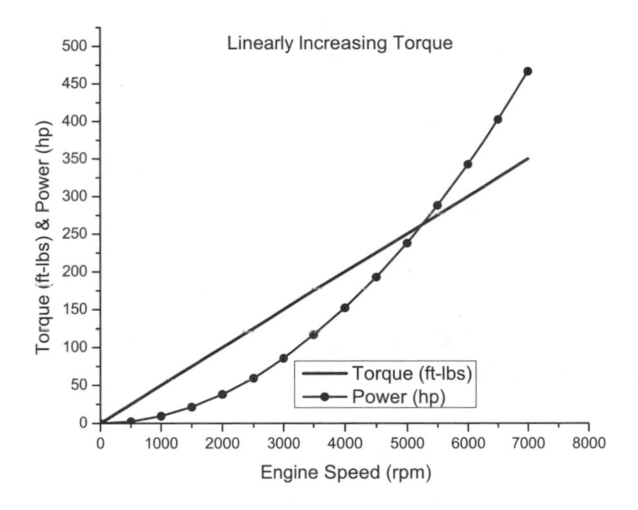

Figure 2.2 Linearly increasing torque causes a concave upward power curve.

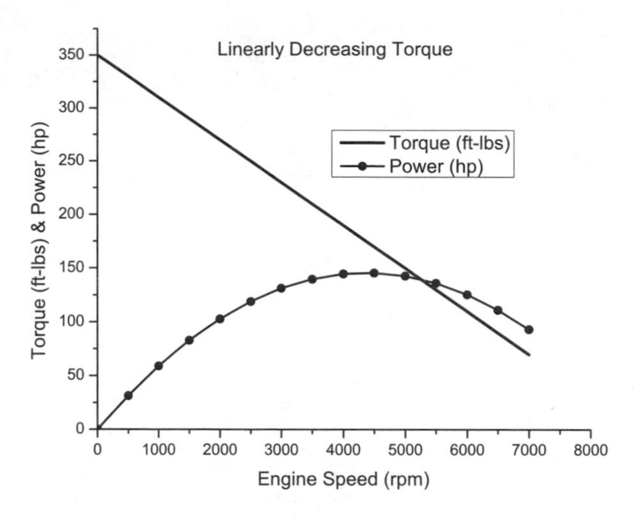

Figure 2.3 For linearly decreasing torque the horsepower is concave downward.

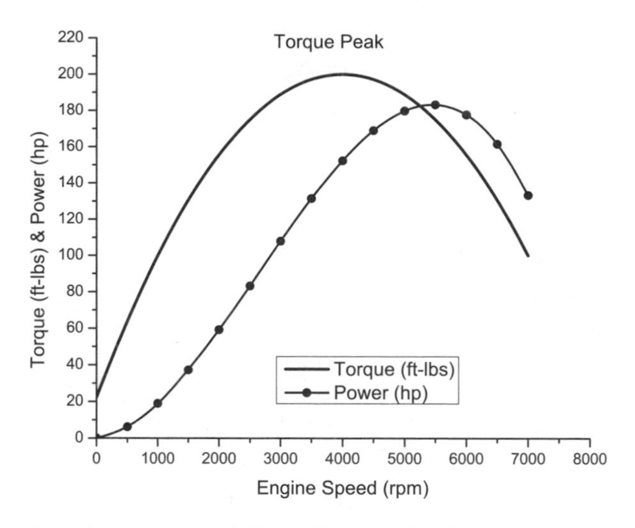

Figure 2.4 A peak in torque is followed by a peak in horsepower.

2.2 HOW DOES DRAG FORCE AFFECT SHIFT POINTS?

To see how the addition of drag force affects the selection of shift points, let's first replot figure 1.6 as the force available at the tire as a function of speed and gear. We do this by multiplying the acceleration by the 102.2 slug mass. The results are shown in figure 2.5.

The drag force is a function of speed. Let's consider what happens at 92 mph. In either fourth or fifth gear, there are approximately 950 lbs of force available at the tire to push the car. Let's say that the drag force is 100 lbs. It will be 100 lbs in either fourth or fifth gear. Therefore, the net force is

$$F_{RT} - f_{\text{drag}} = 950 - 100 = 850 \text{ lbs.}$$

This means that the actual acceleration is less than that predicted by figure 1.6. The drag force has the same effect in both gears; therefore, the shift-point speed or rpm is unaffected. We'll estimate drag and calculate the results in section 2.7.

2.3 GEAR RATIOS

We can learn from figure 1.6 or figure 2.5 why the STi has a 6-speed transmission and not a 4-speed. We can simulate a 4-speed by skipping a couple of gears. I frequently do this in around-town traffic to save wear and tear on my car. If we shift from first gear in the STi at 7000 rpm and go directly to third gear, skipping second, that result is an acceleration of 0.5 g's instead of 0.7 g's. The time to go from 0 to 60 mph would be worse, but because the engine rpm decreases under this scenario, the fuel economy might improve. The Corvette has used just such a strategy in recent years to improve their EPA gas mileage figures.

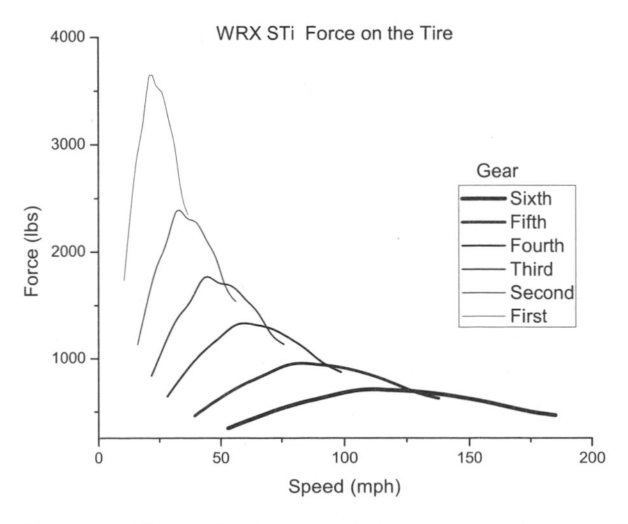

Figure 2.5 Force available at the tire to push the car down the track, F_{RT}.

Having additional gears allows the gear ratios to be closer together. When gear ratios are close, we can achieve a better average acceleration. This effect is slightly offset by two factors. First, additional gears mean a heavier transmission. The second factor is the time it takes to shift gears. When shifting gears, the clutch is disengaged and the acceleration is zero. If you have ever raced away from a stoplight, you know that the first car to shift falls back a little during the brief moment of coasting. A typical value for shifting is one-quarter of a second per shift. We find a great example of taking advantage of the shift times in the comparison of the Mitsubishi Lancer Evolution with the Subaru STi. In 2005 the Evolution had less torque than the STi (273 ft-lbs vs. 300 ft-lbs) but beat the STi in the 0 to 60 mph race (4.8 s vs. 4.9 s). The reason is an extra shift. The Evolution was geared to shift at 40 and 60 mph, while the STi shifted at 37 and 52 mph by our calculations. Even if you try to hold off to the redline in the STi, you must shift at 57 mph. By the time you reach 80 mph, the STi's superior torque and close gear ratios have gained a 0.2 s advantage, and it hangs on to win the 0 to 100 mph and the quarter-mile competition.

Increasing the differential gear ratio will improve acceleration at every speed, but we obtain this advantage at the expense of top speed and probably fuel economy. Engines lose a greater percentage of their power to friction when run at higher rpm. There can be other unintended consequences as well. For instance, in first gear the STi is already at the limit of road tire adhesion. A predicted acceleration of greater than 1 g on road tires can mean a loss of traction (wheel spin, burnouts, or worse) simply as a result of stomping on the gas pedal. It is worth pointing out that at the limit of traction, based on a typical load transfer, a four-wheel-drive car potentially has about 25% more acceleration than a rear-wheel-drive car. We will neglect this for the moment.

2.4 CALCULATING 0 TO 60 MPH TIMES

The maximum acceleration as a function of velocity curves in figure 2.6 makes it easy to understand the effect of torque and the selection of shift points, but it does not directly show the speed and distance as a function of time. To obtain these values, we'll need to review the mathematical relation between acceleration and velocity. The average acceleration is equal to the change in velocity, ΔV, divided by the time it takes to make that change, Δt:

$$a_{\text{average}} = \frac{\Delta V}{\Delta t}. \tag{2.4}$$

The average is over the time interval Δt. A lot can change over that time interval. We can better represent what is going on if we pick a smaller time interval. We can move from an average representation to a real-time or instantaneous representation by considering smaller and smaller time intervals. In the limit where the time interval is infinitesimally small, we have what we defined as the instantaneous acceleration, which is the derivative of velocity with respect to time:

$$a = \frac{dV}{dt}. \tag{2.5}$$

If your calculus is weak (or nonexistent), don't lose heart. We will end up with expressions that can be approached with algebra and a spreadsheet. Equation (2.5) can be rearranged as follows:

$$dt = \frac{dV}{a}. \tag{2.6}$$

If we can express maximum obtainable acceleration as a function of velocity, $a_{max}(V)$, then we can integrate both sides of this equation. Luckily for us, $a_{max}(V)$ can be extracted from figure 1.6. First, we can combine all the data in figure 1.6 into a single curve, which shows the maximum acceleration for any given velocity. This is shown in figure 2.6. For simplicity of notation we will drop the "(V)" and just call it a_{max}.

We must now decide how to deal with accelerations for velocities less than 15 ft/s. Since low-rpm torque is weak, especially in a turbocharged car like the STi, there is a temptation to fill in the gap with a straight line going back to the origin. This would correspond to starting from 0 rpm with the clutch fully engaged and moving off the line. Of course, this is not how a manual transmission is actually used. From a stop, the engine is revved up and the clutch is slipped until the car's speed is large enough to fully engage. In a racing start, this is taken to an extreme, revving the engine near the torque peak and slipping the tires as well as the clutch. The four-wheel drive of the STi makes it next to impossible to slip the tires, so clutch slip is the major mechanism for managing a mismatch between the engine and the transmission speed. The

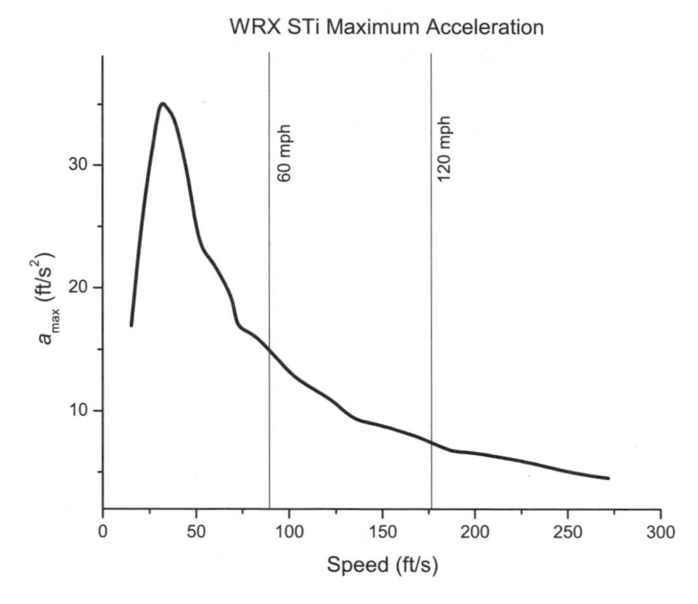

Figure 2.6 Maximum acceleration as a function of velocity.

torque to the wheels will be high, but it cannot equal the maximum because frictional losses in the clutch detract from the available torque. As a first guess, we'll assume that the acceleration is a constant 0.9 g's from 0 to about 23 ft/s. At 23 ft/s the engine rotation rate and transmission rotation rate are both approaching 3000–3500 rpm from above and below, respectively. This value is roughly 80% of the maximum acceleration. Such an approximation might seem critical, but in the end, it corresponds to only about the first 0.8 s of the run. The adapted curve is shown in figure 2.7.

Recall that our goal is to find the time that it takes to achieve 60 mph and that we are setting up the graph that corresponds to the following equation:

$$dt = \left(\frac{1}{a_{max}(V)}\right) dV. \tag{2.7}$$

Using values from figure 2.7, we will plot $1/a_{max}(V)$ as a function of V in figure 2.8. The area under this curve is the time it takes to achieve that particular speed. Ideally, at this point we would integrate equation (2.7) from 0 to 60 mph, as shown in equation (2.8):

$$\int_0^t dt' = \int_0^V \left(\frac{1}{a_{max}(V')} \right) dV'. \tag{2.8}$$

The left side of this equation is easy; it's just the time t. Since we do not have an analytical function, we will need to use a numerical approximation of the area for the right side. The easiest approximation is Euler's method. In Euler's method, the integral on the right is replaced with a summation of rectangles whose area is equal to $[1/a_{max}(V)]\ (\Delta V)$. The $1/a_{max}(V)$ that we will use is the average value over the width of the velocity interval, ΔV. For example, the rectangle on the left in figure 2.8 is 0.1000 s²/ft high and approximately 10.0 ft/s wide for an area of 1.00 s. Equation (2.8) becomes

$$t = \sum_i \frac{1}{a_{max}(V)_i} \Delta V_i. \tag{2.9}$$

To find the time it takes to get to a particular velocity, sum the area of all of the rectangles corresponding to lesser velocities. This sum is easily performed with a spreadsheet.

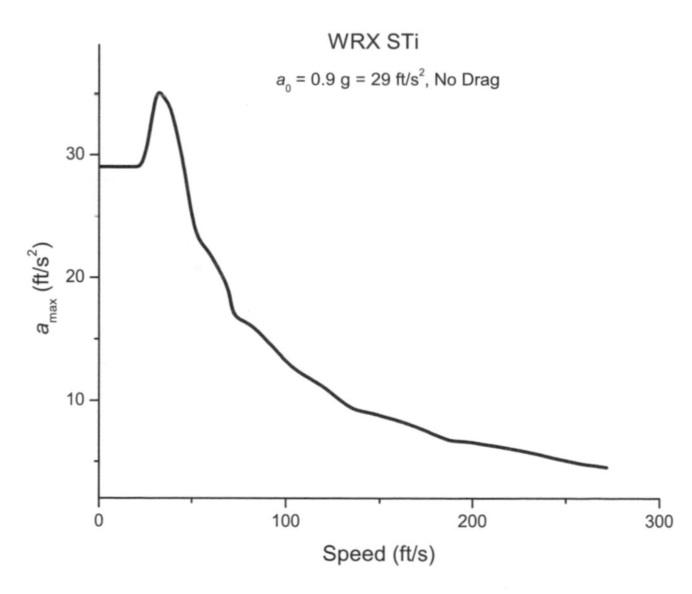

Figure 2.7 Maximum acceleration as a function of velocity with the assumption that an acceleration of 0.9 g's is achieved at low speeds by slipping the clutch.

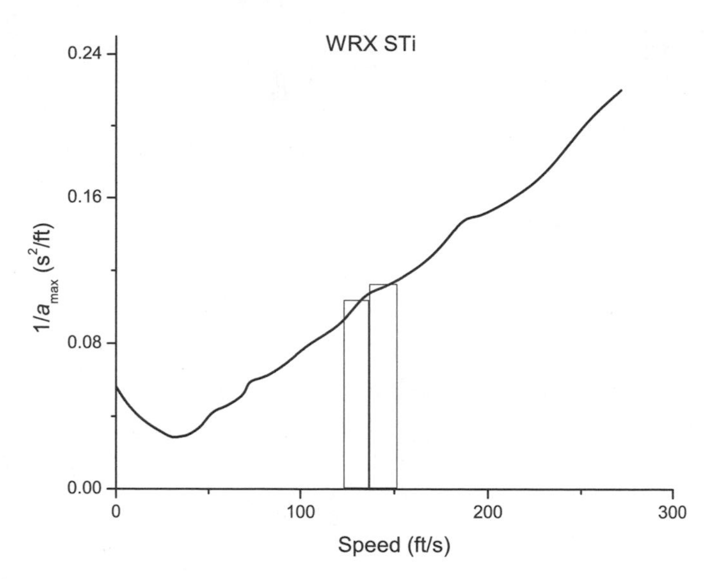

Figure 2.8 $1/a(V)$ as a function of V. The area under the curve corresponds to the time it takes to reach a given speed. The vertical rectangles are examples of the shapes used to approximate the area under the curve in Euler's method.

We need to add to this sum the times that correspond to shifting gears. While shifting, the acceleration is, at best, zero. As a result of drag, rolling friction from the tires, and drivetrain losses, it is in reality slightly negative. We'll use 0.25 s for each shift and assume a zero acceleration for simplicity.

2.5 ASSUMPTIONS AND RESULTS

Here is a summary of our assumptions and results so far:

- Below 3000 rpm in first gear you slip the clutch and maintain 0.9 g's.
- Each shift takes 0.25 s and $a = 0$ during the shift.
- No aerodynamic drag is considered. (We'll come back to this as well.)

The result of the summation is plotted in miles per hour and seconds in figure 2.9, along with reported test results from *Road & Track*. Considering the simplicity of the assumptions, the agreement is amazingly good. The 0 to 60 mph time is only off by 0.6 s. The divergence between the curves at high speed is probably related to drag. We'll discuss that later. The comparison between the *Road & Track* data and the calculated times is summarized in table 2.1.

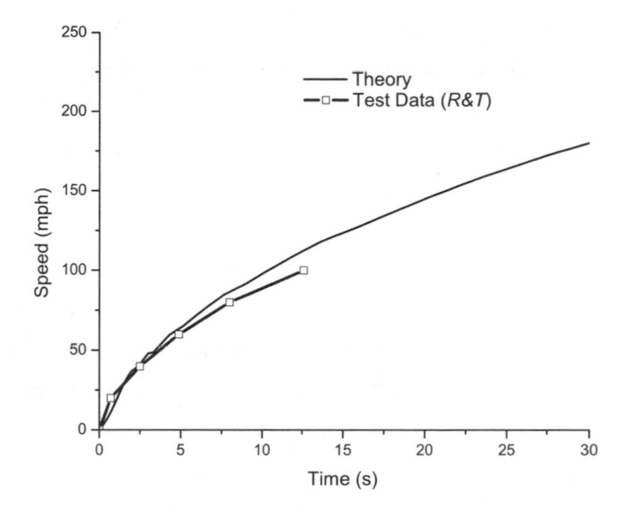

Figure 2.9 The solid line corresponds to predicted speed derived by numerically integrating the area under the curve in figure 2.8. The line with squares is experimental data obtained at the track and reported in *Road & Track Road Test Annual for 2004*.

TABLE 2.1
Experimental and Calculated Time to Various Speeds

Speed Range (mph)	Experimental Time(s) (from *Road & Track*)	Calculated Time(s) with Assumptions
0 to 20	0.7	1
0 to 40	2.5	2.3
0 to 60	4.9	4.3
0 to 80	8.0	7
0 to 100	12.6	10.2

2.6 WHAT IS THE LIMIT FOR 0 TO 60 MPH?

By street car standards, the STi times are fast. Ferrari fast. To reach 60 mph in anything significantly less requires expenditures of $100K or more. The Ferrari Enzo, for example, at $643K with 650 hp and 484 ft-lbs of torque reaches 60 mph in 3.3 s. Is this the limit? Could we estimate the limit? What do we know about the limiting quantities? Torque, gear ratios, shift times, mass, and traction all stand between a car and the limit. Street tires are perhaps the most limiting factor. Rarely do street cars exceed 1.1 g's without switching to racing tires. The limited life span of racing tires makes them completely impractical on the street. The Enzo reaches 60 mph without shifting and has plenty of torque to spare. Assuming that the Enzo is only limited by the 1.1 g's imposed by the tires, we can estimate the minimum time by assuming this constant acceleration.

Constant Acceleration Equations Review

You will recall from general physics that the relationships between acceleration, a, velocity, v, and position, x, for straight-line motion are significantly simplified if we can approximate the acceleration as a constant:

$$x = x_0 + v_0 t + \frac{1}{2}at^2 \qquad (2.10)$$

$$v = v_0 + at \qquad (2.11)$$

$$v^2 = v_0^2 = 2a(x - x_0), \qquad (2.12)$$

where the zero subscript indicates the initial values at time $t = 0$.

Using constant acceleration equation (2.10), from general physics we have

$$\frac{60 \text{ miles}}{\text{hr}} \times \frac{5280 \text{ ft}}{\text{mile}} \times \frac{\text{hr}}{3600\text{s}} = 88\,\frac{\text{ft}}{\text{s}}$$

$$v = v_0 + at$$

$$t = \frac{v}{a} = \frac{88 \text{ ft/s}}{1.1 \text{ g's} \times \dfrac{32 \text{ ft/s}^2}{1.0\text{g}}} = 2.5 \text{ s}.$$

The upper limit on 0 to 60 mph times on street tires is about 2.5 s. Despite the computer-aided launch control used by the Enzo to limit wheel spin, there is still room for improvement.

If your pockets are really deep, you can go for the $1.4 million Bugatti Veyron 16.4. The 16.4 stands for 16 cylinders and four turbo chargers. Its 1001 hp and 922 ft-lbs of torque, when fed through its computer-controlled four-wheel-drive system with launch control, crank out 0 to 60 mph in 2.6 s, all in the comfort of first gear. At about the 10 s mark the Veyron is blowing through 140 mph with no end in sight.

2.7 AERODYNAMIC DRAG

It is common practice in automotive engineering to assume that drag is proportional to speed squared. While this isn't strictly true, it is a good approximation. The typical mathematical expression employed is as follows:

$$F_D = \tfrac{1}{2}\rho A C_D v^2. \tag{2.13}$$

F_D is the drag force in pounds, ρ is the density of air in slugs/ft^3, A is the frontal area of the car in ft^2, C_D is a dimensionless quantity called the drag coefficient, and v is the speed in ft/s. The drag force opposes motion through the air and in our case is in direct opposition to the propulsive force provided by the friction of the tires.

Under what circumstances should we care about drag? Drag is proportional to speed squared, and it would seem that only high-speed vehicles would be concerned. Oddly enough, at times it seems the opposite is true. For example, with a C_D of 1.1 or more, Formula 1 cars have about 4 times the drag coefficient of a Honda Insight (C_D = 0.25) or a Toyota Prius (C_D = 0.27). Similarly, bicycle racers often wear aerodynamic helmets and apply spoke covers to reduce drag, despite the low speed at which they travel. The purpose and limitations of the particular vehicle provide the answer. Yes, Formula 1 cars produce lots of drag, but they also have lots of power. Their designers are more concerned with producing down-force, traction, and stability. They obtain down-force using wings and ground effects at the expense of increased drag. With 800 to 900 hp they are up to the task. The Insight and the Prius, on the other hand, are looking to minimize fuel consumption. Minimizing drag clearly contributes to fuel efficiency, even at low speeds. The bike racer is power and endurance limited. Any reduction in drag force moves the rider further from exhaustion.

It is time to add drag force to our computation and see if we improve our estimate for the STi. Aerodynamic drag is a complex topic. Ferrari employs dozens of engineers and technicians and a wind tunnel running 24 hours a day in support of their Formula 1 racing effort. We'll explore some of the finer points later. Regardless, it is still possible to make a good first-order approximation. Since we do not know the atmospheric details for the day of *Road &*

Track's test run, we will assume that it was at Standard Temperature and Pressure (STP) and thus ρ is 2.38×10^{-3} slugs/ft^3. The frontal area is the total area including the windshield, grill, bumper, and tires. If you backed the car up to a wall and shined a bright but distant light on the car, A would be the surface area of the shadow cast on the wall. We'll estimate 20 ft^2. In terms of the physics, the drag coefficient is where the complexity lies. One of the reasons that modern cars are so fuel efficient compared to the cars of the 1960s is the reduction of the coefficient of drag from the 0.5 region to the 0.3 region found in most non-SUV modern cars. C_D is typically measured in a wind tunnel and is not calculated from first principles. A variety of sources report a value of 0.33 for the STi. Since this is a typical value, we'll go with it. A quick unit check for the right side of the equation yields slug-ft/s^2, which is equivalent to pounds.

Newton's second law now gives us

$$F_{RT} - F_D = ma_{net}, \quad (2.14)$$

where F_{RT} is the force of the road on the tire and F_D is the drag force. Dividing both sides by the mass, we get

$$\frac{F_{RT}}{m} - \frac{F_D}{m} = a_{net}. \quad (2.15)$$

The first term is the acceleration found in figure 2.6. The second term will be the adjustment to the acceleration due to drag. Figure 2.10 is the resulting net acceleration.

Employing the same technique as before, a new graph showing speed as a function of time was generated and is shown in figure 2.11. Clearly, the agreement with published *Road & Track* data is significantly improved. The disagreement below 25 mph is probably due to the way the data are collected using a drag race start. The disagreement above 25 mph is a puzzle. The first temptation is to tinker with the drag coefficient to match the data. This approach quickly fails, since it requires that C_D be more than doubled. The next thought is that some other loss mechanism is required. The dynamometer accounts for most drivetrain losses. The only other widely reported loss mechanism is rolling resistance in the tires. While this loss might be slightly different when running on the dynamometer rollers than it is on the street, it should still show up in the dyno data.

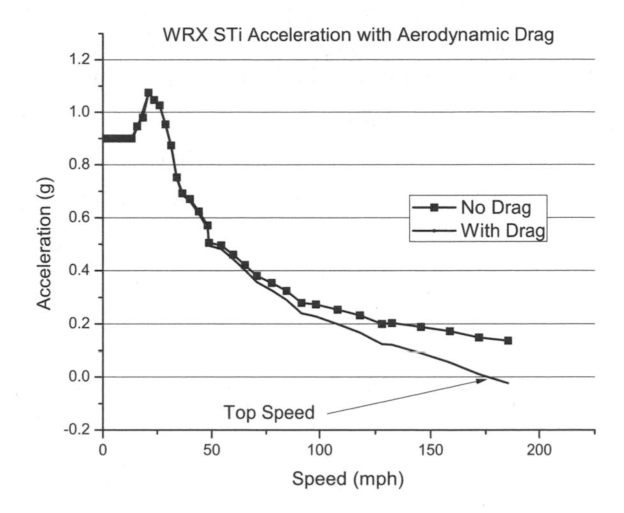

Figure 2.10 Subaru STi acceleration corrected for drag. The reported value of the drag coefficient of 0.33 was employed. Acceleration was assumed to be 0.9 g's for the first 0.9 s.

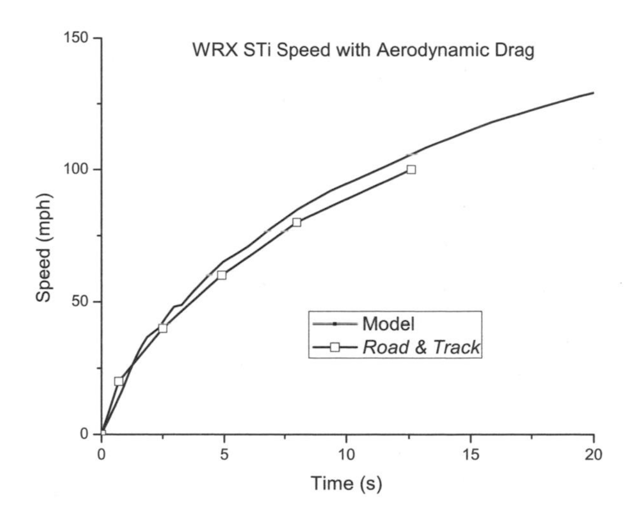

Figure 2.11 Subaru STi speed corrected for a drag coefficient of 0.33.

2.8 CORRECTION FACTORS

A review of a variety of sources of information on dyno data failed to yield an answer until I ran into a couple of Web rants that indicated that some dyno software add unexplained correction factors. A careful review of engineering literature yields a variety of organizations that publish methods for correcting torque and horsepower. DIN (German), ECE (European), and even the individual manufacturers have methods of correction. In fact, Dynojet refers to a 15% "mechanical loss" correction factor. Could this be the source? This reference is to a correction factor developed by the Society of Automotive Engineers, standard J1349. The intent of all of these standards is to ensure repeatability of measurements from day to day. The SAE correction factor, C_f, is as follows:

$$C_f = 1.180\left[\frac{990}{P_d} \cdot \left(\frac{T_c - 273}{298}\right)^{0.5}\right] - 0.18, \tag{2.16}$$

where P_d is the pressure of dry air in hPa (hectopascals) and T_c is the air's temperature in degrees Celsius. The 15% mechanical efficiency is the SAE's estimate of the average power lost by an engine in producing the power. For example, if the engine produces 100 hp, only 85 hp make it to the flywheel. The other 15 hp are lost in friction and in pumping inefficiencies. It is applied as follows:

$$P_{\text{corr}} = \left(\frac{P_{\text{measured}}}{0.85} \cdot C_f\right) - (0.15 \cdot P_{\text{measured}}). \tag{2.17}$$

This means that the correction factor acts on the power made and then the loss term is subtracted out. This is a small correction. The SAE trusts C_f up to a maximum correction of 7%. Beyond that, the equation cannot accurately estimate the horsepower under standard conditions.

What does this mean to us? On any given day, the car performance should be consistent with the dyno-measured values on that day. It will not perform as predicted by the correct values unless conditions are ideal (i.e., pressure is 990 hPa and temperature is 25°C). Since we don't know the conditions of *Road & Track*'s test day, some degree of uncertainty exists. Our dyno curves could be overestimating the car's performance on the track day. Figure 2.12 shows the corrected speed versus time curve with a 7% reduction in dynamometer torque.

Agreement this time is quite good. Does this mean that the dyno data are in fact incorrect? Not necessarily, but it is clear that comparing track and dyno

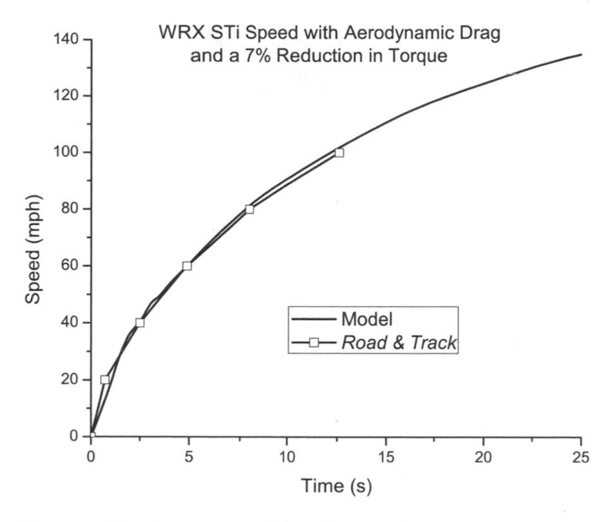

Figure 2.12 STi speed both corrected for drag and removing the 7% dynamometer correction factor.

data requires knowledge of the atmospheric conditions and the correction method. We might have guessed that the *Sport Compact Car* torque data had a significant correction factor applied from the beginning if we had considered drivetrain losses. When compared with the advertised engine torque of 300 ft-lbs, the *Sport Compact Car* chassis dyno value of 267 ft-lbs showed a remarkably optimistic drivetrain torque loss of only 33 ft-lbs or 13%. When added to the 7% correction factor that we guessed, it brings the drivetrain loss to 20%, a much more realistic figure for a complex four-wheel-drive platform.

2.9 THE QUARTER MILE

Now that we have reasonable approximations for velocity as a function of time, we can estimate distance as a function of time and times for the quarter mile. The first step is replotting figure 2.12 with velocity in ft/s. We then apply Euler's method breaking the area under the curve into small rectangles that are v_{i+1} high and Δt wide.

The position, x, is found using

$$x_{i+1} = x_i + v_{i+1}\Delta t, \tag{2.18}$$

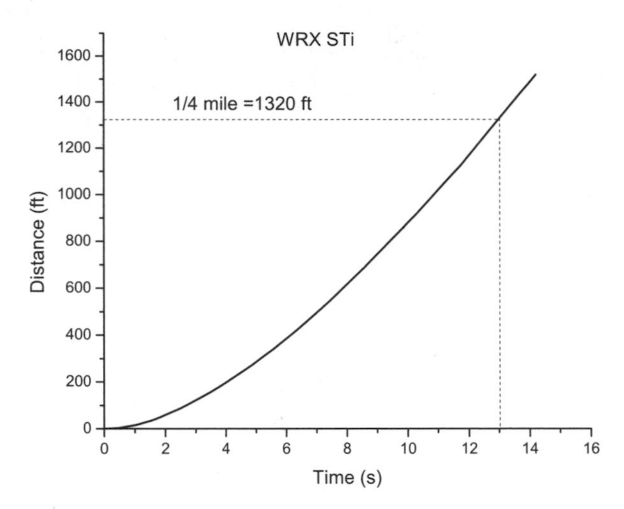

Figure 2.13 Distance as a function of time calculated by applying Euler's method to the curve in figure 2.12. The vertical line is at the quarter-mile distance of 1320 ft and corresponds to 13.0 s. This is 0.3 s quicker than the quarter-mile time reported in *Road & Track*.

where x_i is the sum of all the areas for lesser times. This is another ideal calculation for a spreadsheet. The results are shown in figure 2.13. The quarter mile (1320 ft) is marked by the vertical line.

Within a couple of percent, the prediction in figure 2.13 for the quarter-mile time and values from *Road & Track* agree. This estimate includes the same assumptions as figure 2.12 in that $a = 0.9$ g's for the first 0.9 s, shift times are 0.25 s, and aerodynamic drag has been included. We reduced torque values used by 7% from those reported in *Sport Compact Car.*

2.10 FLAT TORQUE CURVES

Not all sports cars are turbocharged like the STi. Turbocharged and supercharged engines are referred to as having "forced induction" because they use a compressor to increase the oxygen that makes it into the cylinder. The forced induction is responsible, in part, for the large peak in the torque curve. The peak in the torque leads designers to consider the overlap in the acceleration curves that leads to the alternate shift points we found for the STi. We refer to engines without forced induction as "naturally aspirated." Designers of naturally aspirated engines for sports cars frequently attempt to develop engines with flat torque curves. Figure 2.14a is such a set of curves for the Nissan 350Z.

Selecting gear ratios for such a torque curve leaves more options for the designer. Figure 2.14b shows the acceleration versus velocity curves for the

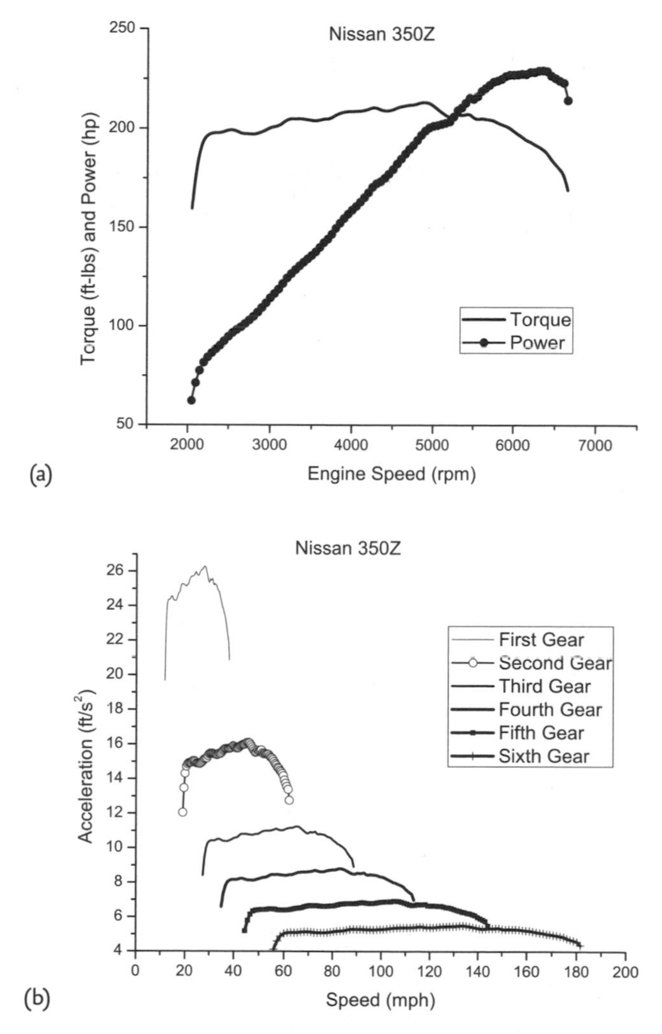

Figure 2.14 (a) Torque and horsepower curves from a chassis dynamometer for a 2003 Nissan 350Z. (b) Predicted accelerations in each gear for the 350Z.

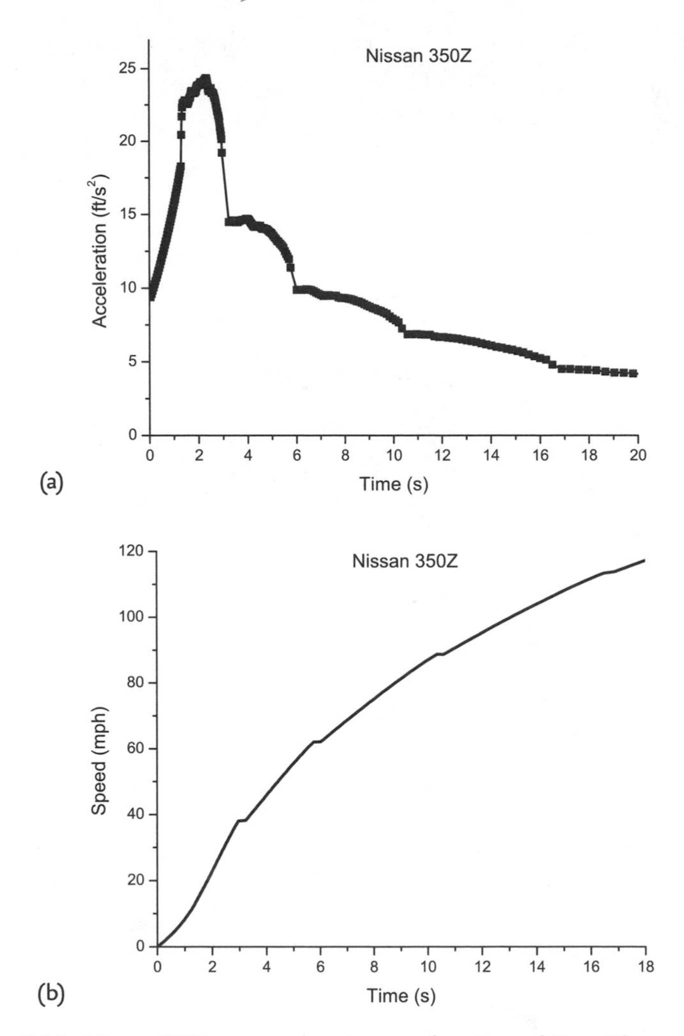

Figure 2.15 Nissan 350Z net acceleration as a function of time. The curve accounts for approximations similar to the STi. One difference is the low-speed acceleration.

350Z as it comes from the factory. Note that the acceleration curves for first to second and second to third never cross. The remainder of the curves meet at the redline. To reiterate, this means that the 350Z should be shifted at redline in every gear.

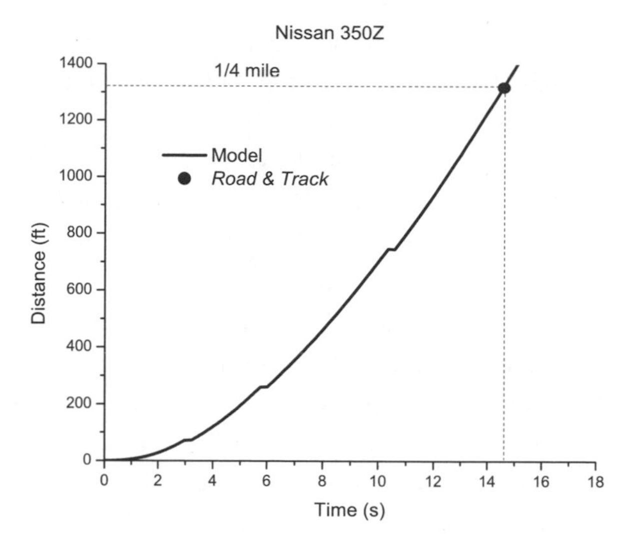

Figure 2.16 Nissan 350Z calculated distance vs. time curves. The vertical line is 1320 ft, the distance for a quarter mile, and reflects a time of about 14.5 s.

The choice of the gear ratios produces 0 to 60 mph times and quarter-mile times that are significantly slower than the STi, despite the fact that the power, torque, and weight are similar. The designers could have improved both the 0 to 60 mph time and the quarter-mile time by raising the second and third gear ratios. Another contributor to the slower times is the fact that the 350Z is two-wheel drive and is unable to launch with the same initial acceleration as the STi. Figures 2.15a, 2.15b, and 2.16 reflect an approximation that the initial acceleration is weaker.

Figures 2.15 and 2.16 include the following approximations:

- Initial acceleration
- Cross-sectional area
- Dynamometer correction factor

2.11 TOP FUEL DRAGSTERS

It is worth taking a moment to consider the performance of the dragsters at the top end of the food chain, the Top Fuel dragsters. As in all drag races, the winner is the first car to cross the finish line. We need to know more about how the race is run to understand the results. At the start line the cars must pull through two beams of light, illuminating a "pre-stage" and a "stage" indicator light. Once the stage light is entered, the yellow and green start lights cycle and the race starts. The elapsed time clock does not start until the car unblocks the stage light beam. The elapsed time (ET) is measured from the time the car leaves the light beam at the start line until it breaks another beam at the finish line, a quarter of a mile away. From the point where the car is properly staged to start the race to the point where it leaves the beam, a Top Fueler can roll up to 18 in. This means that the initial velocity is not zero at the start of the ET. Since the ET clock does not start until you move, you may lose the race and have a better ET than your opponent. The Top Fuel ET record is 4.428 s, set by Tony Schumacher on November 12, 2006. A second measured quantity is top speed, also set by Schumacher on May 25, 2005. Top speed is determined by using a time-of-flight measurement of the last 66 ft of the track before the finish line, called the speed trap. The record is a scorching 336.15 mph!

In June of 2000 *Road & Track* published some velocity and acceleration data for Joe Amato's Top Fuel dragster when it ran a 4.51 s quarter-mile ET

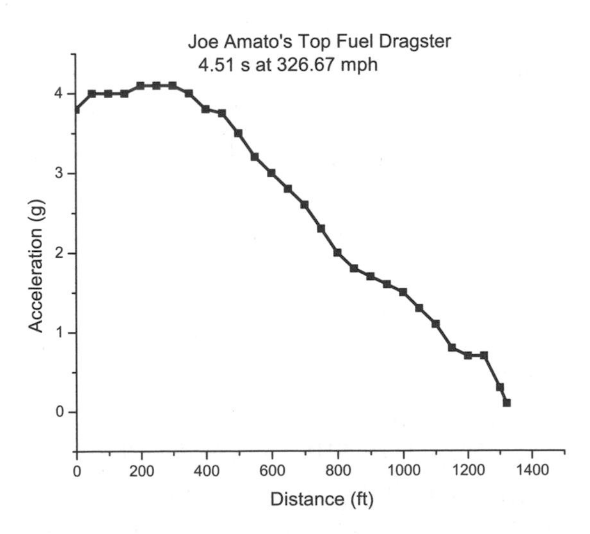

Figure 2.17 Acceleration as a function of distance for a Top Fuel dragster.

TABLE 2.2
Data for Acceleration as a Function of Distance for a Top Fuel Dragster in Figure 2.17

D (ft)	*a* (g)	*a* (ft/s²)	*D* (ft)	*a* (g)	*a* (ft/s²)
0	3.8	122.4	700	2.6	83.7
50	4	128.8	750	2.3	74.1
100	4	128.8	800	2	64.4
150	4	128.8	850	1.8	58.0
200	4.1	132.0	900	1.7	54.7
250	4.1	132.0	950	1.6	51.5
300	4.1	132.0	1000	1.5	48.3
350	4	128.8	1050	1.3	41.9
400	3.8	122.4	1100	1.1	35.4
450	3.75	120.8	1150	0.8	25.8
500	3.5	112.7	1200	0.7	22.5
550	3.2	103.0	1250	0.7	22.5
600	3	96.6	1300	0.3	9.7
650	2.8	90.2	1321	0.1	3.2

with a trap speed of 326.67 mph. While this is not quite record pace, it does give us some insight into Top Fuel performance. Acceleration as a function of distance is shown in figure 2.17, and the data are summarized in table 2.2.

The first thing you notice about figure 2.17 is that the acceleration for the first 400 ft is greater than 4 g's. This means it feels like a force of more than 800 lbs slams a 200 lb driver into his seat! We could have guessed this by calculating the average acceleration: 326 mph is roughly 479 ft/s; thus, the average acceleration is

$$a_{avg} = \frac{\Delta v}{\Delta t} = \frac{(479 - 0)}{(4.51 - 0)} = 106.2 \text{ ft/s}^2 = 3.3\text{g's}.$$

What can we extract from figure 2.17? The slope, $\Delta a/\Delta x$, doesn't ring a bell. How about the area under the curve? We can break the area into rectangles of height $a(x_i)$ times the width Δx. If we multiply this product by the mass of the dragster, we have the force times the displacement, which is the work done on the dragster:

$$ma(x_i)\Delta x = F_i\Delta x = \Delta W, \tag{2.19}$$

where F_i is the force and ΔW is an element of work done over the displacement Δx. If we sum over all the rectangles under the curve and multiply by the mass, we get the total work done on the dragster. The work–kinetic energy theorem tells us that the net work done on an object is equal to its change in kinetic energy, as shown in equation (2.20):

$$\sum_i W_i = \Delta KE = \frac{1}{2}mv_f^2 - \frac{1}{2}mv_0^2. \tag{2.20}$$

Kinetic energy is the energy associated with the motion of an object and is equal to $\frac{1}{2}mv^2$, where m is the mass and v is the speed of the object. The car has a weight of 2150 lbs, which corresponds to 66.8 slugs. Following our previous unit conversions, the area under the curve must be converted to ft-lbs. These units are equivalent to the units on the right side of equation (2.19) and (2.20), which are slug-ft^2/s^2. The area times the mass of the car is equal to 7.38×10^6 ft-lbs. The initial velocity for this Top Fuel car, after rollout, is 9 mph. Solving equation (2.20) for the final speed yields 470 ft/s or 321 mph. With less than 2% difference from the reported trap speed, this result is pretty good. Note that the work-energy approach tells us nothing about the time it takes to make the quarter-mile run. We need to do a Newton's second law calculation to get time information.

They typically don't make dynamometers that work in such a high power range, so most quoted values are estimates. So, let's make an estimate. Power is work per time:

$$\text{Average Power} = (7.38 \times 10^6 \text{ ft-lbs})/(4.51 \text{ s}) = 1.64 \times 10^6 \text{ ft-lbs/s}$$
$$\text{or roughly 3000 hp.}$$

We can look at instantaneous power. The instantaneous power of a force that is collinear with the velocity is equal to the magnitude of the force times the magnitude of the speed. For the first 350 ft the acceleration is approximately 4 g's. We can apply a constant acceleration equation over this range.

The net instantaneous power is

$$v^2 = v_0^2 - 2a\Delta x \text{ which leads to } v = 300 \text{ ft/s}$$
$$P = Fv = mav = (66.8 \text{ slugs})(4(32.2))(300) = 2.58 \times 10^6 \text{ ft} - \text{lbs/s}$$
$$P = 4690 \text{ hp.}$$

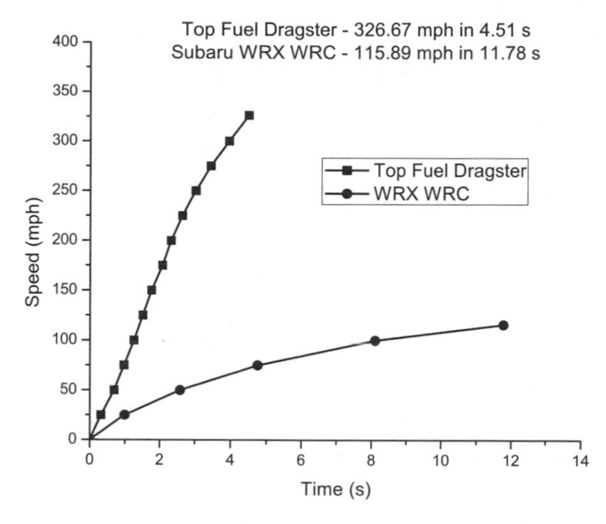

Figure 2.18 Speed as a function of time for a Top Fuel dragster and World Rally Championship STi. Source of data: *Road & Track,* June 2000.

Power is positive for the engine's contribution and negative for the aerodynamic drag, since the force opposes the velocity. The car must overcome the drag force. From equation (2.13), $F_D = \frac{1}{2}\rho AC_D v^2$. We'll use $C_D = 0.5$, $A = 20\ \text{ft}^2$, and $v = 300$ ft/s. The magnitude of this power is

$$P_D = F_D V = \tfrac{1}{2}\rho AC_D v^3$$

$$P_D = \frac{1}{2}(2.38\times10^{-3})(20)(0.5)(300)^3$$

$$P_D = 3.21 \times 10^5\ \text{ft} - \text{lbs/s}$$

$$P_D = 584\ \text{hp}.$$

The total power produced by the car is the sum of net power plus the aerodynamic drag power, for a grand total of 5274 hp. A 25% drivetrain loss puts our estimate at roughly 7000 hp.

Finally, figure 2.18 shows a comparison between Joe Amato's Top Fuel dragster and a World Rally Championship STi. While this STi is fast by street standards, the screaming Top Fuel rocket dwarfs its performance.

2.12 SUMMARY

We have explored the relationship between torque and horsepower, estimated 0 to 60 mph times and quarter-mile times from torque curves, estimated the best 0 to 60 mph time for a street car, and, finally, gained an appreciation for the performance of Top Fuel dragsters.

Lessons learned:

- We can convert dynamometer data into approximate performance values. Dynamometer data must be properly calibrated, as must any track-day test values. Low-speed (<10 mph) acceleration must be estimated.
- Given a torque versus rpm curve, we can easily estimate the shape of the horsepower curve.
- At 5252 rpm, the number for horsepower is equal to the number for torque in ft-lbs.
- The peak torque value is an indicator of acceleration. Actual performance requires knowledge of the entire torque curve.
- Peak horsepower is an indicator of top speed.
- Street tires are capable of a minimum 0 to 60 mph time of about 2.5 s, and getting to this level of performance isn't cheap.
- Top Fuel dragsters run the quarter mile at an estimated peak of 7000 hp, at 4 g's of acceleration, more than 320 mph, and an elapsed time of approximately 4.5 s.

Chapter 3

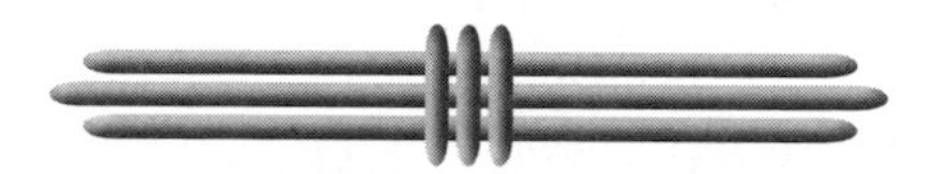

Finding the Racing Line: Road Racing

In road racing, the tracks look like sections of public rural roads. The asphalt or concrete turns both left and right. The radius of the corners varies, as do the elevation and camber of the roadway. A well-built track will feature large runoff areas where cars that lose control can scrub off speed without collisions. A typical track will cover 1.5 to 3.5 miles with 10 to 15 different turns. For a driver to be competitive, he or she must find the fastest path around the track with the car constantly operating at the limit of acceleration and traction. We call this path the racing line. Under racing conditions, the driver will deviate from this path to pass another car or to avoid being passed, but the driver's goal is to quickly return to the racing line. On a simple track, you can learn the line by following a fast driver around the course during practice. Of course, each car is a little different from the next. What works for a Porsche 911 GT3 does not work for a Honda Civic Si. In order to adapt, a driver must understand the principles that lead to the fastest path.

It's tempting to start the learning process with one of the great road racing circuits, such as the Nürburgring, also simply called "the Ring." This German track has existed in various forms since the early 1920s. Elite Formula 1 racing uses the modern version of the Ring. At 3.2 miles in length with 16 turns, it is a classic modern circuit. However, when road race fans discuss the Ring, it is the Nordschleife (northern loop) that fills their daydreams. This section of the original track, shown in figure 3.1, still exists. This 20.8 km (12.9 mile) loop with 72 turns is open to the public, for a modest fee. The track, lined with trees and featuring dozens of blind turns, has claimed 78 racers' lives. Jackie Stewart of Formula 1 fame nicknamed it the "Green Hell." Many automotive manufacturers, including BMW, still use it as the yardstick by which their cars are measured. For the mere sum of 19 euros, you can take your own car for a lap. If transporting your car to Germany is a bit steep for your budget, you can pick out any one of a number of video gaming systems and battle your way through a virtual lap of this historic circuit. Either way, a hundred laps is considered enough to begin learning the racing line.

For our purposes, the Nordschleife seems a bit over the top. Oval racing is a significant simplification. Oval racing refers to a class of tracks that includes everything that resembles a distorted circle. The Indianapolis Motor Speedway, one of the most famous, is actually a 2.5 mile rectangle with four rounded corners. Each corner produces a 90° direction change. Each lap consists of four left turns. Banking of the roadway allows the 1530 lb, 650 hp, ethanol-burning Indy cars to lap at an average of 225 mph. Aerodynamics and banking play a big role at Indy. This is still very complex. As is true in all physics, we want to start with the simple and work toward the complex. The track at Indy also contains a road race circuit that uses part of the oval and a complex infield section with a total of 2.6 miles with 13 turns. Turn 1 of this course is our ideal starting place. It is a flat 90° right-hand turn. Formula 1 cars drop to a speed of 70 mph and generate

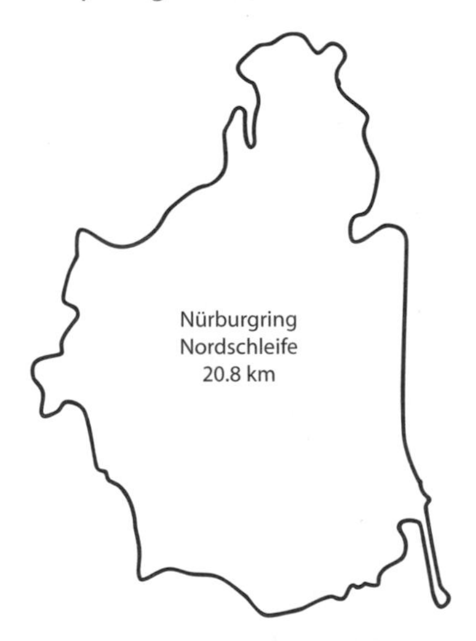

Figure 3.1 The Nürburgring Nordschleife (northern loop). At 20.8 km, it is considered the most challenging road race circuit in the world.

about 3.3 g's of lateral acceleration in turn 1. We will use the 90° flat turn as the cornerstone for our initial modeling of the racing line. To model the physics with changing directions, we'll need to expand our use of vectors to two dimensions. We will start with a quick review.

Two-Dimensional Vector Review

Vectors are physical quantities that have both a size, called the magnitude, and a direction. Force is a vector quantity. For example, you can push your car with a force of 100 lbs northeast across the parking lot. We typically represent a vector with an arrow in drawings. The length represents the magnitude, and the direction of the arrow shows the direction in which the force acts. Bold characters such as the ***F*** shown in the figure below are vector quantities. Non-boldface characters represent scalars, which are quantities that don't have a direction. An example of a scalar would be something like temperature. A vector quantity with the bold removed, such as *F,* represents the magnitude of the vector. Compass directions are more natural for some, but we prefer to use a mathematical coordinate system. In the figure below the east-west direction will become the *x*-direction and north-south the *y*-direction.

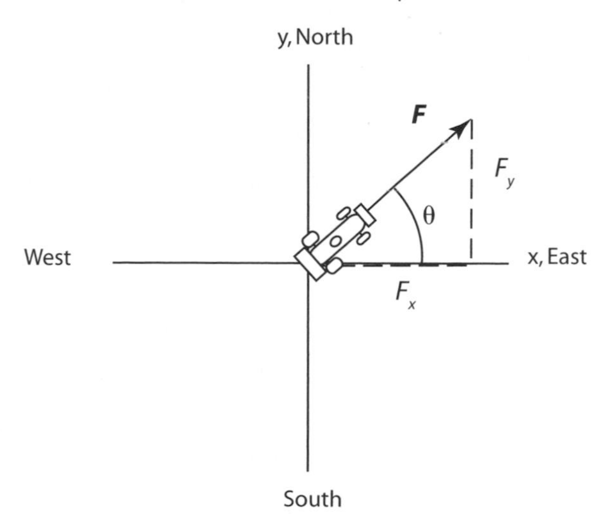

We can make a right triangle by dropping a perpendicular line from the tip of the arrow to the *x*-axis. The sides of the triangle are called components and are shown here as F_x and F_y. For right triangles, Pythagoras taught us that

$$F^2 = F_x^2 + F_y^2.$$

The vector ***F*** makes an angle θ with the *x*-axis. We will find some basic trigonometry useful. In trig sine, cosine, and tangent are nothing more than the ratios of the lengths of the sides of our right triangle. For any particular value of θ, these ratios are well known and can be estimated by most calculators and spreadsheets. If we know the magnitude and the angle, we can find the components with trig.

From trigonometry, we recall that

$$\sin(\theta) \equiv \frac{\text{Length of the opposite side}}{\text{Length of the hypotenuse}} = \frac{F_y}{F} \text{ or } F_y = F\sin(\theta)$$

$$\cos(\theta) \equiv \frac{\text{Length of the adjacent side}}{\text{Length of the hypotenuse}} = \frac{F_x}{F} \text{ or } F_x = F\cos(\theta)$$

$$\tan(\theta) \equiv \frac{\text{Length of the opposite side}}{\text{Length of the adjacent side}} = \frac{F_y}{F_x}.$$

If we know the lengths of the sides, we can find the angle using the inverse trig functions arcsine (written $\sin^{-1}$), arccosine (written $\cos^{-1}$), and arctangent (written $\tan^{-1}$):

$$\sin^{-1}\left(\frac{F_y}{F}\right) = \theta, \qquad \cos^{-1}\left(\frac{F_x}{F}\right) = \theta, \qquad \tan^{-1}\left(\frac{F_y}{F_x}\right) = \theta.$$

Again, the inverse values are well known and can be easily applied using a calculator or a spreadsheet.

3.1 THE TRACTION CIRCLE

For most driving situations, four general external forces contribute to the acceleration of a car: gravity, the normal force of the ground pushing upward on the tires, any aerodynamic force, and friction between the tires and the road. These four forces determine the net acceleration of the car. As we did in chapter 1, we shall begin by simplifying the problem with a few assumptions. First, let's assume that our car operates in the speed regime where we can neglect aerodynamic forces, including drag and the aerodynamic down-force.

Secondly, we will consider a flat track with a uniform surface. This means that gravity and the normal force produce no net acceleration and that the road surface provides a constant coefficient of friction. Finally, let's assume that the maximum frictional force between the tire and the road is isotropic in the plane of the road surface. This means that the tire is capable of producing the same magnitude of frictional force in any direction as long as it is tangent to the surface of the road. These simplifications are significant and neglect many important effects that we will add back in as we develop our understanding.

Our vector quantities in chapter 1 acted directly along our direction of motion, either adding to or opposing that motion. In this section, our friction force will act both along our direction of motion and perpendicular to it. A common tool for analyzing a driver's effective use of the available friction is called the traction circle or the friction circle. For any combination of speeding up, slowing down, or turning, we can draw a friction vector, ***f***, responsible for that acceleration as shown in figure 3.2. The *x*-component in the figure

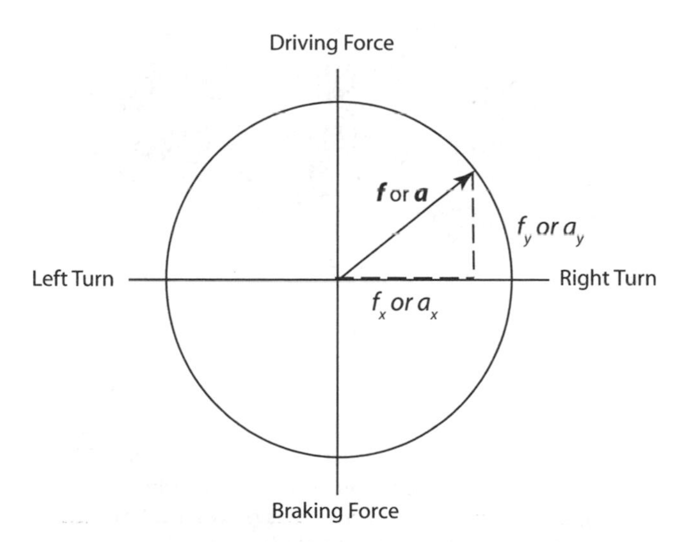

Figure 3.2 The friction force, ***f***, may be broken into lateral and longitudinal components, f_x and f_y, that act to turn and speed up (or slow down) the car at the same time. The circle represents the maximum friction available. At speeds where aerodynamic forces are small, an identical-looking plot can be obtained by plotting the acceleration. The acceleration is also referred to as a g-g diagram.

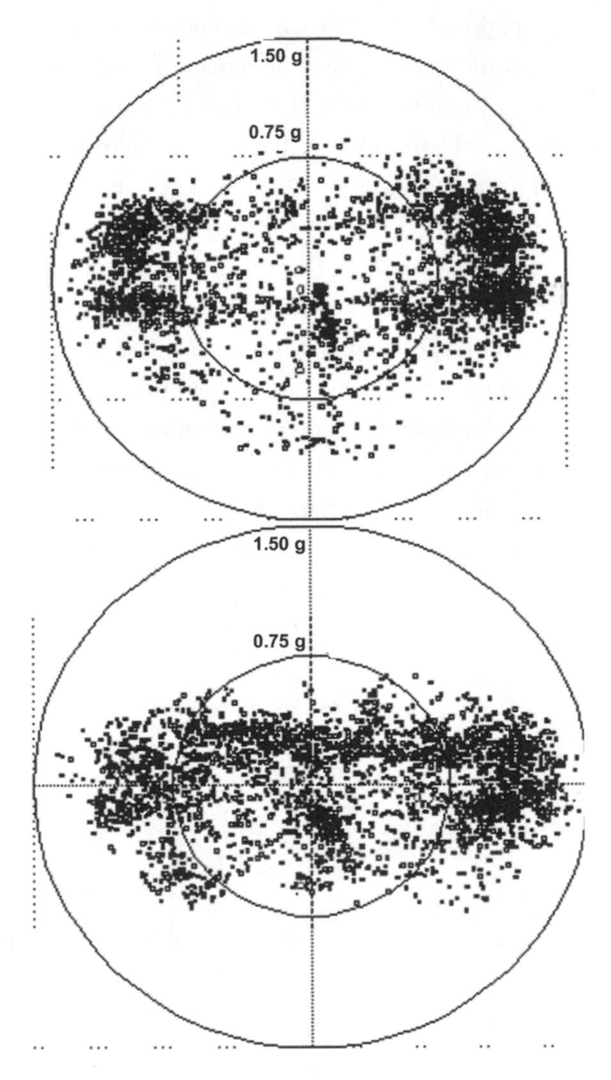

Figure 3.3 These two plots for different drivers show typical friction circle diagrams from automotive data acquisition systems. The axes are in units of g's of acceleration, leading to its other typical name, the g-g diagram. The horizontal axis represents lateral acceleration and the vertical axis longitudinal acceleration. Which driver brakes harder? Do both drivers turn equally well to both the left and the right? Which driver simultaneously turns and brakes with a greater acceleration?

contributes to turning left or right, and the y-component increases or decreases the speed. If the magnitude of the friction corresponds to the maximum static frictional force, then we can draw a circle about the origin with a radius equal to the maximum friction magnitude. For our ideal car, in order to lap the track at the highest speed, in the shortest time, our driver must strive to produce a friction force vector that is near to the surface of this circle.

If we divide the magnitude of the friction by the mass of the car, then, from Newton's second law, we will have a similar diagram for the acceleration vector. This type of plot is called a g-g diagram, since the units for the acceleration are typically represented in g's, similar to our approach in chapter 1.

Figure 3.3 shows a couple of g-g diagrams generated by a two-axis accelerometer mounted in a sports car. Each point corresponds to the acceleration, in units of g's. Data are recorded at equal time intervals during a drive.

The drivers did a good job keeping the lateral acceleration at or above the 0.75 g value but had trouble with longitudinal acceleration. In this case, the drivers had a decent excuse, given that his racing slicks had trouble dealing with the light rain that was falling. This implies that there is a difference between the way the tire handles lateral and longitudinal acceleration.

The g-g diagram, as we have defined it, represents the acceleration for an inertial reference frame. An inertial frame is one in which the coordinated system itself is not accelerating. When this is true, it is easy to follow Newton's laws. Objects at rest in a reference frame do not move unless some net external force acts on them. Place this book on your tabletop, and it stays there until a net force acts on it. What happens to the book when the reference frame itself accelerates? What if your book and table were in a motor home and the driver were to slam on the brakes? The book would fly forward off the table without any force acting on the book. It moves in the frame of the motor home because the motor home accelerated at some rate **A**. Newton's second law is corrected by adding a new term:

$$\mathbf{F}_{\text{net}} - m\mathbf{A} = m\mathbf{a}.$$

The second term $m\mathbf{A}$ is sometimes referred to as the inertial term. The net force acting on an accelerometer bolted to the floor of a racing car is zero in the reference frame of the car. It stays where you bolted it down. When the car accelerates, the accelerometer feels the inertial term. Step on the gas, and it "feels" like it's being pushed backward. Stomp on the brakes, and it "feels" like it's being thrown forward. Turn hard left, and it "feels" like it is thrown to the right. The accelerometer records the inertial term, which has the opposite sign of the acceleration of the car when viewed from the reference frame of the road. Some g-g diagrams in other sources will plot the data using the sign convention of the inertial term. Braking will be in the positive y-direction and speeding up in the negative y-direction. A left turn will be plotted on the positive x side of the graph. It can be confusing until you determine the sign convention used by the information source.

Before we move on with our simplified ideal model, it is worth considering some of the factors that alter the shape of the traction circle. An ideal street tire is capable of producing about 1 g of acceleration in all directions. Our corresponding maximum friction would produce a circle. We found in chapter 1 that the Subaru STi produced a g only in first gear. With each shift of the gears, the acceleration fell further. Road racers use first gear at the start and in the pit, but rarely on the track. This means that the traction circle for the STi will be less by about half a g in the plus y-direction. Aerodynamic downforce at high speeds will increase the normal force and in turn expand the maximum friction circle. This effect can be significant. In Formula 1 racing, the cars can brake and turn at 4 to 5 g's at high speed! As the car slows, that maximum friction drops to below 2 g's. To know if the driver is making maximum use of the available traction, you'll have to know their speed. All tires, but especially racing tires, have a coefficient of friction that is temperature dependent. When the tires are cold, their traction is poor. Wrecks during the first lap after a tire change are common. Some road racing series allow the use of tire warmers, which are essentially electric blankets. On the other hand, when tires are too hot, their traction is also poor. This is typically the result of overdriving the car, scrubbing the tires across the pavement at every turn or braking too late. Thus, the temperatures of the tires alter the shape of the maximum friction circle.

An increase in the vertical load and normal force linearly increases the available friction force, to a point. Beyond that limit, the coefficient of friction decreases with increasing load. We will discuss this further in chapter 4. Clearly, the friction force generated by deformable surfaces is more complicated than the ideal rigid bodies studied in physics class. Clearly, tire loading can alter the friction diagram. By design, a drag racing tire produces lots of longitudinal force, but very little lateral force. Suspension design can alter the orientation of the tire with respect to the road as well as the rate of load transfer. Thus, tire and suspension design can also alter the shape of the maximums on the friction diagram. The fact that this problem is complex will not prevent us from gaining an understanding of the contributing factors, if we consider them one at a time.

3.2 NINETY DEGREE RIGHT-HAND TURN

Our goal here is to gather enough information to construct a model of a simple rectangular racetrack. We can then create a simple spreadsheet and calculate lap times. This will allow us to change parameters, such as acceleration or straightaway length, so that we can determine their importance. We'll start with a simple right-hand turn.

Figure 3.4 shows our basic 90° right-hand turn. Let's pick an inner radius,

Clyde Caplan looks far ahead on the autocross track, planning his approach for each turn. Photograph by Clyde Caplan and Alex Teitelbaum.

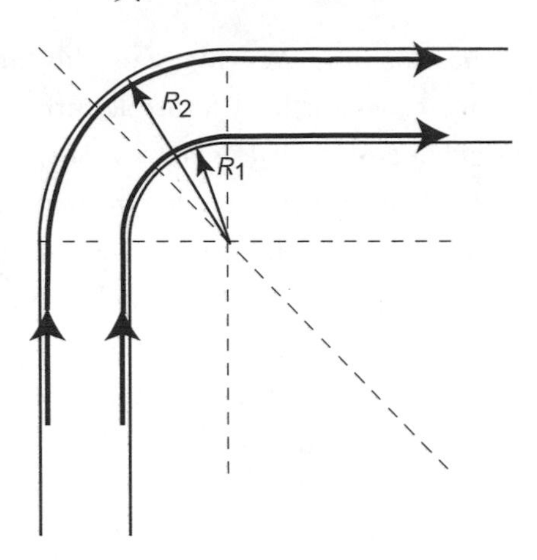

Figure 3.4 Racing lines hugging the inside or outside edge of the track.

R_1, of 36 ft and an outer radius, R_2, of 78 ft. We are going to consider the corner in isolation. We'll make some assumptions about the capabilities of our car and analyze several different paths.

The car is initially 600 ft from the start of the turn and traveling at a constant 124 mph. The traction circle tells us that the car is most effective at braking in a straight line. Both mathematically and physically, turning while braking is challenging. Therefore, for starters, let's assume that all the braking is complete before we turn. We'll assume that we are on street tires and that the coefficient of friction, μ, is equal to 1. This allows all accelerations to be 1 g. It is reasonable to brake and turn on street tires at 1 g. Speeding up at 1 g is overly optimistic for a street machine, but reasonable for a starting model. As usual, we'll start with a simple example and increase complexity later. We will follow the path that hugs the inside edge of the track, and for simplicity we'll ignore the fact that the car has width to it. When do we brake? That depends on how much we need to slow down. The slowest speed is limited by the rate at which we can corner. We start by considering the corner.

During the turn, the car travels at constant speed. We call this uniform circular motion. The free-body diagram for this situation is shown in figure 3.5. From this, we can write Newton's second law by components in the *x*- and *y*-directions. The vertical, or *y*-direction, is in equilibrium between the weight, *mg*, and the normal force, *n*, between the ground and the car. For

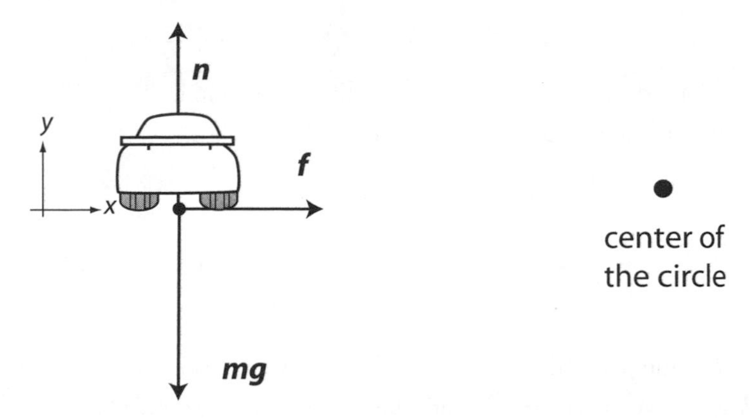

Figure 3.5 Free-body diagram for a car traveling in uniform circular motion. We have treated the car as a particle; therefore, all forces act at a single point.

example, the sum of the force in the x-direction is equal to the mass times the acceleration in the x-direction, and mathematically

$$\sum F_x = ma_x \qquad \sum F_y = ma_y$$
$$f = ma_x \qquad n - mg = 0$$

From the y equation, for our flat, horizontal road we see that the normal force is equal to the weight. Since we don't want the tires to skid, we are dealing with static friction. Substituting, we have

$$f_s = \mu_s n = \mu_s mg. \tag{3.1}$$

For an object to travel in a uniform circular path, the centripetal acceleration must be equal to V^2/R, where V is the tangential speed and R is the radius of the circle. Substituting this into the second law, we have

$$f = ma_x \tag{3.2}$$

$$\mu_s mg = \frac{mV^2}{R} \tag{3.3}$$

$$V = \sqrt{\mu_s Rg}\,. \tag{3.4}$$

The speed is proportional to the square root of the radius of the turn. Following the inside radius of the turn,

$$V = \sqrt{(1)\,(36\text{ ft})\,(32.2\text{ ft/s}^2)} = 34.0\text{ ft/s or } 23.2\text{ mph}.$$

How long does it take to drive around the quarter circle? Since the speed is constant, we can use a simple expression:

$$V = D/t \qquad \text{(distance/time)}$$

$$t = D/V = \frac{\tfrac{1}{4}[2\pi R]}{\sqrt{\mu_s R g}} = \left(\frac{\pi}{2\sqrt{\mu_s g}}\right)\sqrt{R}.$$

For our inside path this yields $t = 1.66$ s. Next we consider the constant braking section. We need to calculate the braking distance and the time under braking. We start braking at 124 mph and end at 23.2 mph, or 182 ft/s down to 34.0 ft/s. We'll use the constant acceleration equations again. How far will we go under braking?

$$V_F^2 = V_0^2 + 2a(x_F - x_0)$$

$$(x_F - x_0) = \frac{-182^2 + 34.0^2}{2(-32.2\text{ ft/s}^2)} = 496\text{ ft}.$$

How long will it take?

$$V_F = V_0 + at$$

$$t = \frac{34.0\text{ ft/s} - 182\text{ ft/s}}{-32.2\text{ ft/s}^2} = 4.60\text{ s}.$$

The section before that is at a constant 124 mph. The distance at this speed is the 600 ft straightaway length minus the 496 ft braking distance, yielding 104 ft. Thus,

$$t = \frac{D}{V} = \frac{104\text{ ft}}{182\text{ ft/s}} = 0.57\text{ s}.$$

As with all physics models, we will start with simplifying assumptions. If the speeding up, $\boldsymbol{a}_{up}$, following the turn is equal and opposite to the braking section, $\boldsymbol{a}_{down}$, the two times will be the same. This also applies for the final constant velocity section. The total time is

$$T_{total} = 0.57 \text{ s} + 4.60 \text{ s} + 1.66 \text{ s} + 4.60 \text{ s} + 0.57 \text{ s} = 12.0 \text{ s}.$$

Repeat this calculation for the driver that follows the outside edge of the corner. The time comes out nearly identical. The straightaway times are less because the driver can enter the corner with more speed. The time gained on the two straightaways is lost in the corner because the path is longer. It is common to follow a path like this the first time a driver gets to drive on a track. It is the way we drive on the street, staying in our lane and avoiding the traffic around us. It takes a while to break these habits and take advantage of the entire track surface.

To improve our time through the corner, we must shorten our path and carry more speed. The path that accomplishes this is shown in figure 3.6.

A path that maximizes speed and minimizes path length is the circular arc of radius R_3. It is tangent to the outside edges of the track on the entry and exit straights and is tangent to the inside edge of the track at the midpoint of the corner. The inside edge contact point is called the apex of the corner. The center of curvature for R_3 still lies on the axis of symmetry, but it lies a distance, d, beyond the center of curvature for R_1 and R_2. As a result, the curve for R_3 extends a distance, x, beyond the boundaries of the original turn. This geometry defines a right triangle shown in figure 3.6b. The angle θ is 45°.

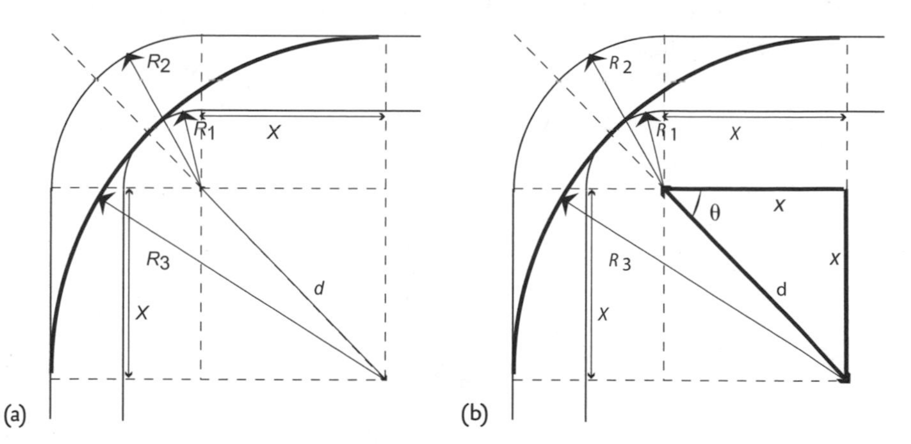

Figure 3.6 (a) Symmetrical turn with a mid-corner apex. (b) Right triangle used to analyze the 90° right turn.

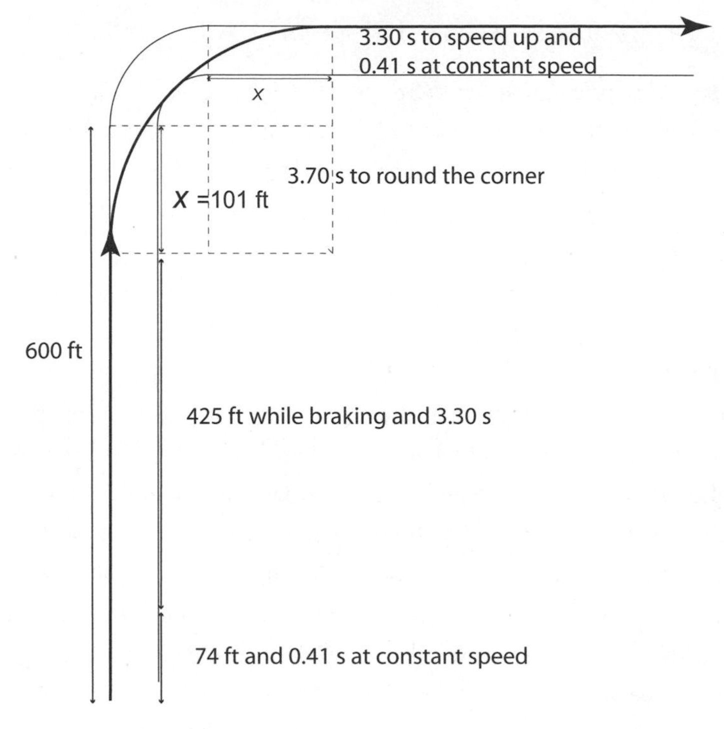

Figure 3.7 Times for the 90° mid-corner apex turn.

The goal is to find an expression for R_3 in terms of the known quantities. A little staring at the figure reveals three relationships:

$$R_1 + d = R_3$$
$$R_2 + x = R_3$$
$$x/d = \cos\theta.$$

Combining the three equations and solving for R_3,

$$R_3 = R_1 + \frac{1}{1 - \cos(\theta)}(R_2 - R_1)$$
$$R_3 = 179 \text{ ft.} \tag{3.5}$$

The speed on the R_3 path is

$$V_3 = \sqrt{\mu_s g R_3} = \sqrt{(1)(32.2)(179)} = 75.9 \text{ ft/s or } 51.8 \text{ mph.}$$

The time for this path is

$$t = \frac{D}{V} = \left(\frac{\pi}{2\sqrt{\mu_s g}}\right)\sqrt{R} = 3.70 \text{ s.}$$

The R_3 path takes away a distance x from the straightaway length:

$$x = R_3 - R_2 = 179 - 78 = 101 \text{ ft.}$$

This leaves the straightaway with 499 ft. During the braking section, the car slows from 182 ft/s to 75.9 ft/s at −1 g. This takes 3.30 s and covers 425 of the 499 ft. This leaves 74 ft at 182 ft/s, which takes 0.41 s. These are shown in figure 3.7.

The total time for the R_3 path is

$$t = 3.70 \text{ s} + (2\cdot 3.30 \text{ s}) + (2\cdot 0.41 \text{ s}) = 11.1 \text{ s.}$$

This is a 0.9 s advantage over our result for just following the curve of the corner. At 124 mph, 0.9 s corresponds to a lead of 164 ft, or about nine car lengths!

3.3 GENERAL TURN

Equation (3.5) is valid for any angle change as long as the turn is symmetric. The θ in this equation is half of the direction change, just as it was in our example. Figure 3.8 shows a general symmetric turn with a direction change of 2θ. Can you see the similar right triangles that could be used to prove equation (3.5)? Using the same road width as our previous example and a 60° direction change, the resulting R_3 path has a radius of 349 ft and a speed of 106 ft/s, or 72.3 mph.

3.4 CONSTRUCTING A TRACK MODEL

We now have enough information to construct a model of a simple rectangular racetrack and a spreadsheet to calculate lap times. Laying out the spreadsheet such that track and car parameters are defined in a single cell will make it easier to study the effect of changing those parameters.

In figure 3.9 we have a rectangular track with two different paths. We will follow the same approximations as we did for the isolated corner. All speed-

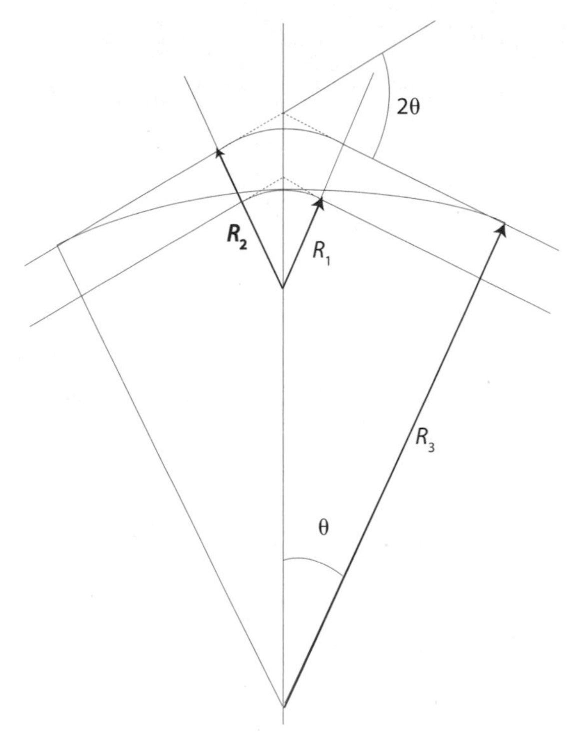

Figure 3.8 This general turn changes direction by 2θ. Equation (3.5) is still valid for finding the radius, R_3, of the high-speed path through the corner.

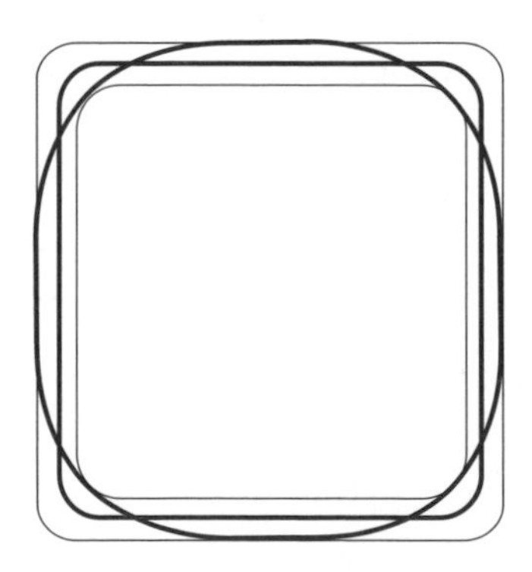

Figure 3.9 Idealized square track with two different possible paths.

ing up or slowing down will be done on the straight sections. The coefficient of friction will remain 1. With 600 ft straightaways the square path represents a half-mile track. If the drivetrain can produce 1 g of acceleration, a lap of the square path takes roughly 33.5 s. The path with the high-speed corners takes about 30 s.

For a production car, as we saw in chapter 1, 1 g is a large acceleration for speeding up. The engine and drivetrain simply do not produce enough torque. Figure 3.10 is a plot of

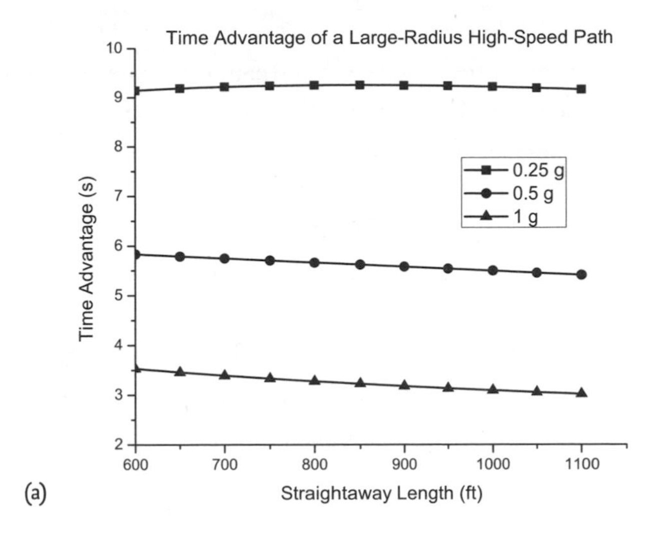

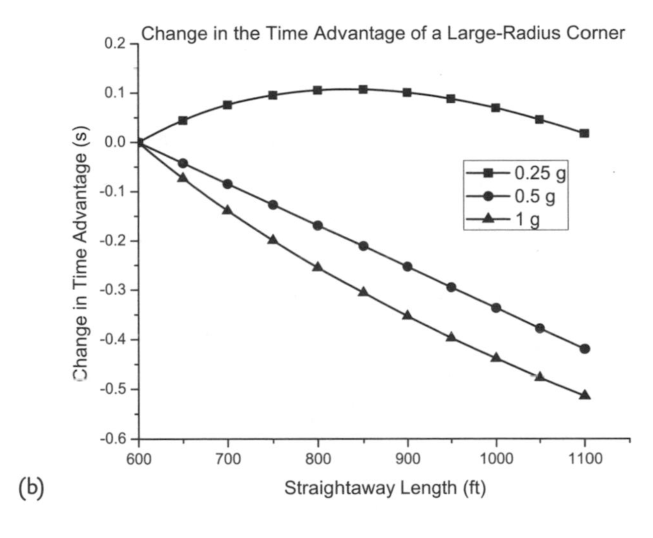

Figure 3.10 (a) Time advantage of using large-radius turns for an entire lap of the track in fig. 3.9 for three different accelerations while speeding up, a_u. The lower the acceleration, the more significant is the advantage. The time advantage, $t(L)$, is a function of the length of the straightaway, L. It is difficult to judge the degree of change in this plot. (b) Plot of the change in the time advantage, Δt, from fig. 3.10a. For example, when $a_u = g/4$, the time advantage for a 900 ft straightaway is $t(L = 900 \text{ ft}) = 9.24$ s and for a 600 ft straightaway is $t(L = 600 \text{ ft}) = 9.14$ s and $\Delta t = 0.1$ s.

the change in the time advantage between the fast and the slow path for one lap, as a function of the length of the straight. It assumes that our ability to speed up is power limited and considers three accelerations, 1 g, 0.5 g, and 0.25 g. It assumes that the braking is traction limited and assumes 1 g acceleration for slowing down.

The lowest powered car benefited the most (about 9 s) from following the high-speed line and continues to gain time until the straight is about 850 ft long. However, the most intriguing thing about this plot is that the high-power car (the one with 1 g of acceleration and a 3.5 s gain) saw its high-speed path advantage fade as the straight grew longer. Why? The answer comes in the next section and goes by the name "exit speed."

3.5 TYPES OF TURNS

Limiting our thinking to optimizing a single corner misses the importance of how successive parts of the track relate. In figure 3.11, for example, turn 1 is followed immediately by turn 2. Based on our previous work, we should enter turn 1 at point A and exit at point B. Similarly, turn 2 would call for an entry at point C and an exit at point D. We can't be at both B and C at the same time. Which turn is more important, turn 1 or turn 2? Or should we compromise and just drive in between B and C?

In 1971 Alan Johnson, a four-time national champion in the Sports Car Club of America, wrote the book *Driving in Competition*, with advice for the

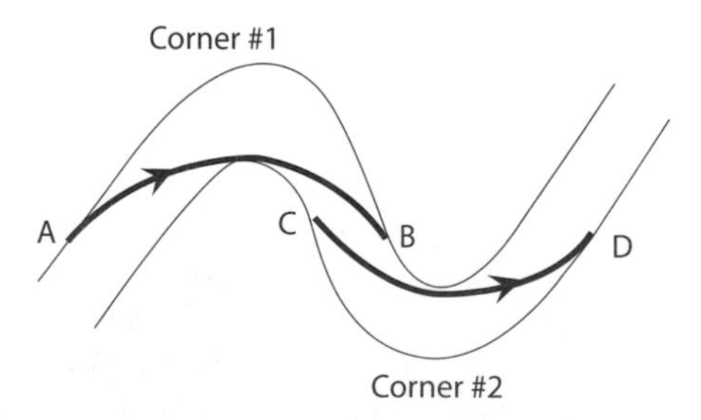

Figure 3.11 The thick black lines represent the quick path through corners 1 and 2. Since the ending point of turn 1 at B is on the opposite side of the track from the starting point for turn 2 at C, we have a problem. We need more information to optimize our driving line.

novice racer. He addressed the complexity of finding the racing line by dividing turns into three fundamental types. Nearly 40 years later, these categories are still used by driving instructors across the country. Our goal here is to use some basics physics to validate some of his empirical observations.

His first observation was that races are won and lost on the straightaway and not in the turns. The most important aspect of driving through turns is how they help you attain success on the straights. Type 1 turns are the most important. They are the ones at the entrance to a straightaway. Type 2 turns are the ones at the end of straightaways, and type 3 turns are the transition turns between type 1 and type 2 turns.

When two equal cars exit a corner, the one with the highest exit speed will win the battle down a long straight, even if they fall behind a little to attain the greater exit speed. Type 1 turns are most important because they help you attain this exit speed.

For example, consider two identical cars side by side exiting a corner where one car is 5 mph faster than the other at the corner exit. If both can produce the same constant acceleration, a, we can write the following equations:

$$x_S = x_0 + v_S t + \frac{1}{2}at^2$$

$$x_F = x_0 + v_F t + \frac{1}{2}at^2 ,$$

where x is position, v is velocity, and t is the time. The S subscript is for the slower car and F for the faster car. Parameter x_0 is the initial position of both cars at the corner exit. How far apart are they after some time period t? Combine the two equations:

$$x_F - x_S = (v_F - v_S)t.$$

If the velocity difference is 5.00 mph, or 7.30 ft/s, and we wait 5.0 s, we have

$$\Delta x = (7.3 \text{ ft/s})5.0 \text{ s} = 36.5 \text{ ft},$$

or roughly one and a half car lengths. The longer the straightaway following the turn is, the greater the final separation between the cars.

The question then becomes, how do we increase our exit speed? Alan Johnson's answer is to give up some corner entry speed and set up for a late apex

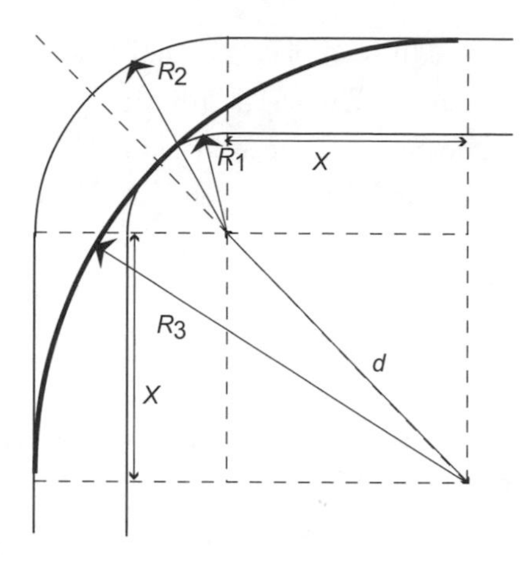

Figure 3.12 Optimum symmetrical cornering line.

exit. By slowing more on corner entry, we can tighten our initial cornering radius and finish the majority of the turning early. This gives us the option, late in the turn, to open the radius and accelerate much sooner. The early acceleration leads to higher exit speed. The mathematics of simultaneous turning and acceleration is more complex. First, let's see what we can learn from turning at constant speed.

In our discussion of optimizing an individual symmetrical turn, we increased the radius of the turn to carry the highest possible speed through the turn. This is shown in figure 3.12. We moved the center of curvature down the axis of the turn and adjusted the apex until it just touched the inside edge of the corner at the midpoint. We started and ended the turn at the outside edges of the track. All of our friction force was devoted to turning, and our entry speed was equal to our exit speed. In this symmetrical case, no higher cornering speed was attainable.

3.6 TYPE 1 TURN

A type 1 turn is shown in figure 3.13. The initial turn of entry radius, r_{en}, is tighter than R_3 of our generic right-hand turn and therefore slower at speed V_{en}. The direction change is quick but ends early. At transition point t, the driver opens the radius to R_{ex}, which is greater than R_3. The available traction will support a much greater speed, V_{ex}. The driver squeezes the throttle, balancing the combined radial and tangential accelerations. The increased radius of curvature results in an apex that happens after the corner midpoint and is referred to as a "late apex." If the tangential acceleration is great enough, the exit speed will be greater than that of a driver on the R_3 curve. This is the origin of the racing phrase "Slow in, fast out." Some think that the phrase means taking it easier on the entry to improve control. In fact, it involves driving at the limit and making a continuous smooth transition to achieve an improved exit speed.

How much additional speed is possible? It depends on the exit path, which

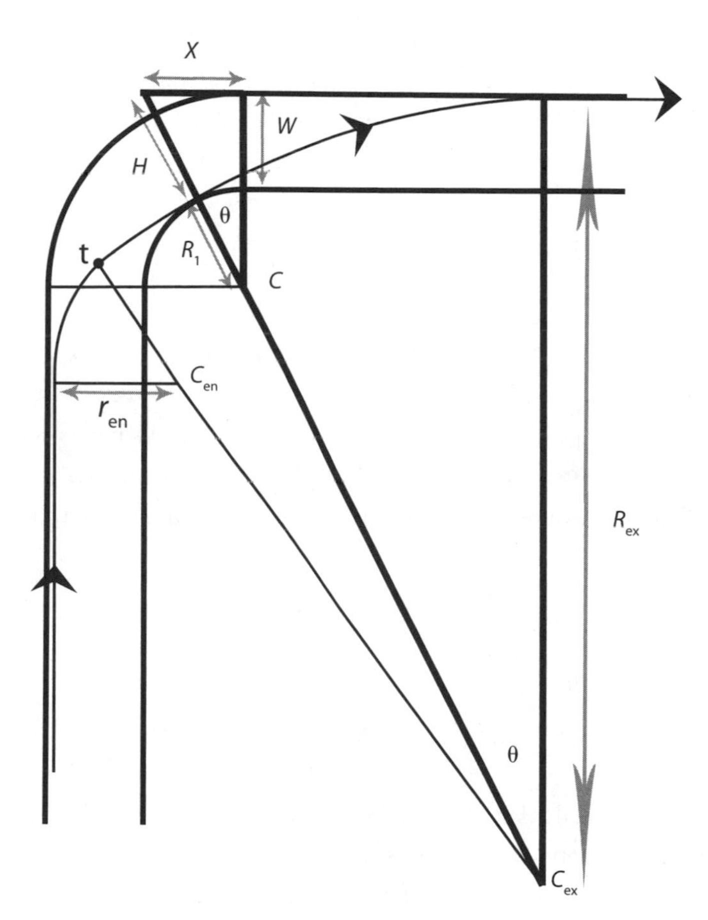

Figure 3.13 Type 1 turn entering a straightaway.

we can characterize in terms of the location of the apex. Figure 3.13 defines some of the geometry for our 90° right-hand bend. C_{ex} is the center of curvature for the corner exit path. A line from the apex to C_{ex} forms an angle θ with a vertical line on the figure. If we extend the line through the apex by a distance H, we can form a right triangle. From the figure we can see that

$$\cos(\theta) = \frac{R_{ex}}{(R_{ex} + H)}.$$

A second similar triangle is formed starting from the center of curvature, C, for the corner. For this second triangle we have

$$\cos(\theta) = \frac{(R_1 + W)}{(R_1 + H)},$$

where W is the width of the track, or $(R_2 - R_1)$. Applying these two cosine relations, we can express the exit radius in terms of the angle θ:

$$R_{ex} = \left(\frac{W}{(1 - \cos(\theta))} + R_1\right). \tag{3.6}$$

If the apex is at the midpoint of the corner, then θ is 45°. If the apex is at $\phi = 5°$ past the midpoint, then $\theta = 45° - \phi$. Given the angle φ, we can find the exit radius, $R_{ex,}$ and from that we can find the maximum exit speed, V_{ex}, on this path. At this speed, all of the traction is devoted to centripetal acceleration and the driver must, once again, hold the throttle steady. Applying Newton's second law, our expression for centripetal acceleration, and assuming a flat track, we have

$$\sum F = f = \mu mg \quad \text{and} \quad \sum F = ma = m\frac{V^2}{R}$$

or (3.7)

$$V_{ex} = \sqrt{\mu g R_{ex}}.$$

We can now calculate our percent speed advantage over the mid-corner apex line as a function of φ. This is shown in figure 3.14.

We continue to refer to the exit speed calculated by equation (3.7) and plotted in figure 3.14 as the "possible speed." Several factors can prevent us from

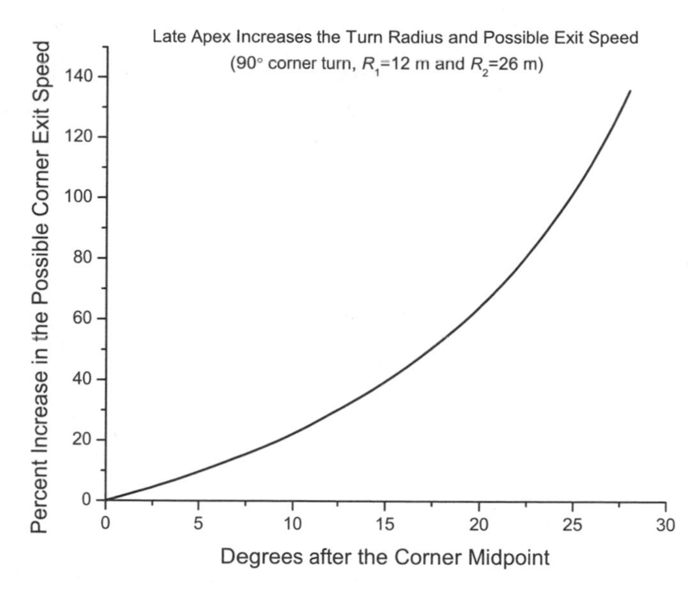

Figure 3.14 Using the type 1 turn strategy shown in fig. 3.13 to attack a 90° right-hand bend improves the possible exit speed over that attained using a corner with a midpoint apex.

attaining this limit. For example, if the driver fails to speed up aggressively to reach V_{ex} and turn at the limit, they may have a lower exit speed. Since the car has already given up speed on corner entry and then exited at a lower speed, the situation is twice as bad. It may sound simple to avoid, but believe me it is challenging. The driver must squeeze on the throttle at the transition point (labeled "t" in fig. 3.13) until the car is speeding up and turning at the limit. As the car increases speed, more traction must be devoted to turning and the centripetal force. The driver must ease off the throttle once the limit is attained and continue to do so until the corner is exited.

To make the math easier, our description of the problem is a little contrived. In reality, as the driver reaches the transition point and squeezes on the throttle, weight transfer and the tire slip angle conspire to open the radius without significant motion of the steering wheel. In effect, the driver uses the throttle to push the car to the outside of the turn. Even under these conditions, it is difficult to keep the car at its traction limit. We will discuss the details of weight transfer and tire slip angle in the following chapters.

It may have occurred to you that a low-power car with limited acceleration may not be able to regain the speed lost on corner entry. For example, consider a car capable of speeding up (tangential acceleration) at 0.2 g's that gives up 10 mph on corner entry. How long will it take to regain that lost speed? We can apply the constant acceleration equations again to make this estimate:

$$v = v_0 + at$$

$$t = \frac{v - v_0}{a} = 2.28\text{s}.$$

This car requires 2.28 s to recover the lost entry speed. Such a car would be forced to increase the entry radius and entry speed, leaving less opportunity to accelerate on the exit of a type 1 turn. These cars are sometimes referred to as "momentum cars" because of their need to maintain speed throughout the corner. In a misuse of physics, drivers talk about "maintaining momentum through the corner." Momentum is the product of mass times velocity, a vector quantity. As shown in figure 3.15, there is a significant change in the vector momentum. However, it is changed by applying a centripetal force from the tires, perpendicular to the direction of motion. Strictly speaking, it is the magnitude of momentum that is maintained throughout the turn. In contrast, a high-power car gives up some of the magnitude of momentum on entry, changes direction, and restores the magnitude using the engine. The racing

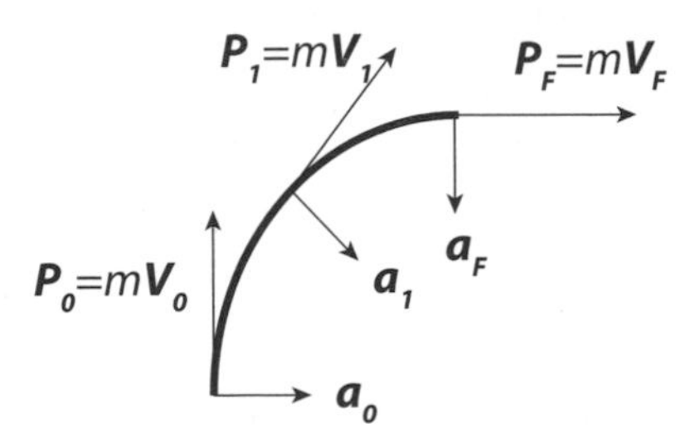

Figure 3.15 A constant-magnitude momentum is maintained when the acceleration remains perpendicular to the motion. When viewed from an inertial frame of reference, the acceleration is centripetal or directed toward the center of curvature. When viewed from the car, it is referred to as lateral acceleration. In either case, the source of the acceleration is friction between the tires and the road.

lines for these two types of cars are slightly different. Later we will address some of the other factors, like understeer and oversteer, that alter the optimal line through the corner.

It is important to point out that the late apex used in a type 1 turn is the safest approach to a corner. Speed is slow on entry, giving the driver more time to make corrections and more asphalt to use before hitting the low-traction dirt or gravel that usually follows. The high-speed part of the turn arrives when the car is already parallel to the shoulder. This is the easiest condition to recover from, should the driver drop a wheel off of the pavement and onto the dirt.

3.7 TYPE 2 TURN

The next most important turn is the type 2 at the end of a straightaway. The strategy is still to maintain the highest speed for the longest time. As shown in figure 3.16, we can achieve this with a straight entry and an early apex, the opposite of a type 1 turn. The driver must head deep into the turn before brak-

Six-time SCCA Solo National Champion Sam Strano nails an early apex driving the Strano Performance Parts/Capital Quest Mortgage 2007 Shelby GT Mustang owned by Mike Snyder. Photograph by Clyde Caplan and Alex Teitelbaum.

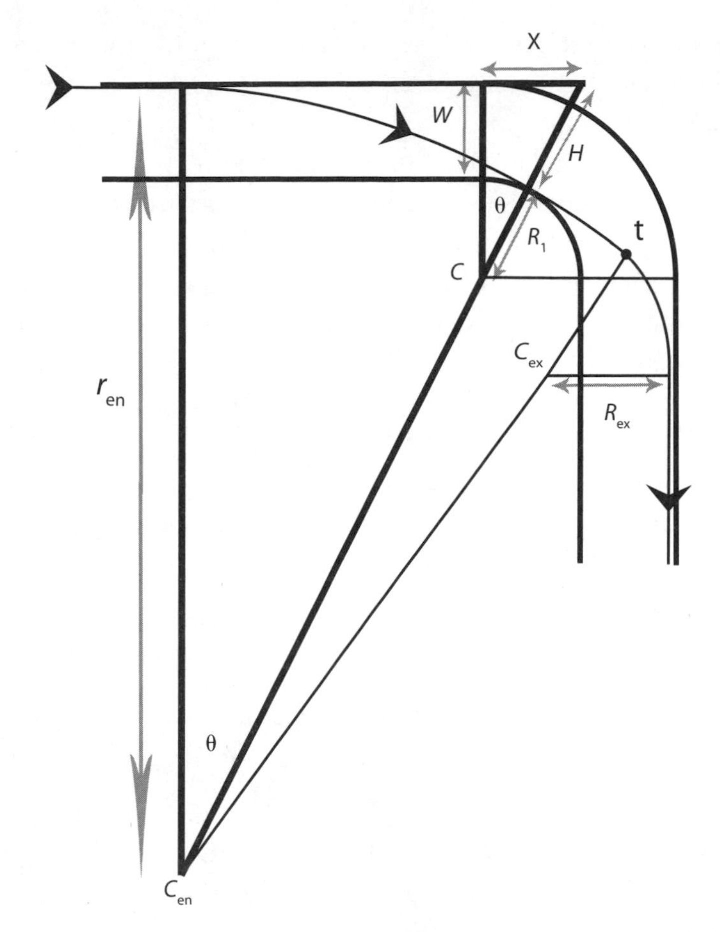

Figure 3.16 A type 2 turn is found at the end of a long straightaway. It is characterized by an early apex and late braking, allowing the driver to maximize the time at high speed on the straight.

ing. A wide radius on entry allows heavy braking during a gentle initial turn. As the speed drops late in the turn, the driver eases off the brakes and tightens the radius at the transition point, t, to complete the turn.

The time advantage of a type 2 turn over a mid-apex turn is significantly less than that of a type 1 turn. The exit speed from a type 2 is extremely slow. If this turn is a transition between two straights, this low-speed exit is compounded over the following straightaway. Remember, as we showed in the previous section, a car that exits with a lower speed than a competitor's will lose ground over the entire straightaway. *Therefore, if a turn is both a type 1 and a type 2, always treat it as a type 1.*

3.8 TYPE 3 TURN

> In SCCA Solo racing, the track is constructed with pylons and is therefore different for every event (one of the appealing aspects of the sport for many). The straights are short and the turns are plentiful. Because of the changing track and limited number of runs to make corrections, the pre-race walk-through becomes an important part of the sport. People walk the track to build a virtual map in their heads, look for visual cues, and begin to construct the racing line. When I started the sport, I would see people go through the turns and, at the exit of the turn, turn around and look at the corner from the other direction. I thought this was dumb. The course is only run in one direction, so what could you learn from looking at it backwards? However, when I look at it from the perspective of the three types of turns, this behavior makes sense. Seeing the turn from the other side may help provide a clearer understanding of where the driver needs to exit the turn to maximize acceleration or set up for the turn that follows.
>
> —Tristan Edmondson

A type 3 turn is one that connects two other turns. Johnson calls it "the least important type of turn on the race track." Because this category encompasses such a wide range of possibilities, it is difficult to describe a concise strategy. When the opportunity presents itself, keep the path straight, minimize the distance, and keep the speed high. As we saw in our initial analysis of the 90° turn, keeping the speed high and the path short are often at odds with each other. Keeping the speed high often means using the entire track width from shoulder to shoulder. Keeping the path short means hugging the inside edge of a 90° corner. A type 3 turn does not have the benefit of a follow-on straightaway to use the additional exit speed. Don't add any distance to the path that does not pay off with enough additional speed to immediately move you ahead of a car on the shorter path. Smooth applications of the steering wheel, brakes, and throttle will keep the car settled and will optimize the use of available

Lee Piccione exits a type 3 turn in Kirk Boston's B Stock 350Z, planning ahead by looking far in advance. Photograph by Clyde Caplan and Alex Teitelbaum.

traction. Above all, drive the track in a way that properly sets up the approach to the type 1 turn that follows. The easiest mistake to make is to overdrive type 3 turns in such a way that you are out of position or traveling too fast to execute the type 1 turn. Consider the S-shaped turns of figure 3.17.

In figure 3.17 we start by trying to apply a symmetric high-speed line through the corners when considered individually. We originally asked, how do we resolve the discrepancy between the exit of the first turn and the entry of the second turn? We now have tools to establish the general line. When considered in the context of the overall track, the second turn is the most important turn, a type 1 turn. To maximize our exit speed at point H, we must set up a late apex in the second turn. With this late apex we can begin to apply the throttle and speed up starting at point G. To accomplish this, we must sacrifice speed and the line in the first turn. We must be slow enough at point F to turn tightly and hug the right-hand edge of the track. By starting the first turn, the type 3 turn, at point E, we can increase the radius of the turn at point F and carry a little more speed and still reach the proper entry to the type 1 turn. It may seem a little strange to analyze the course backward, but the needs of the type 1 come first.

Many other factors may require adjustments to driving line. For example, the camber of the road may help or hinder you while turning. If the road tilts toward the center of curvature, called a "banked turn," you can carry more speed. Banked to a staggering 31°, Daytona International Speedway is home to some of America's most important races. NASCAR routinely cracks the 200 mph barrier around its banked tri-oval corners.

If the road is "off-camber" or tilts away from the center of curvature, you

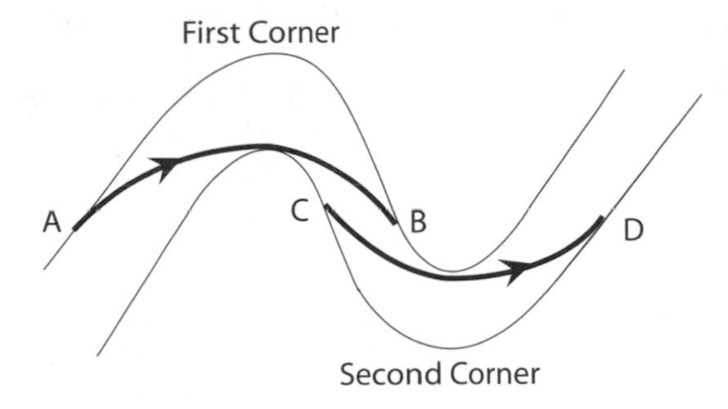

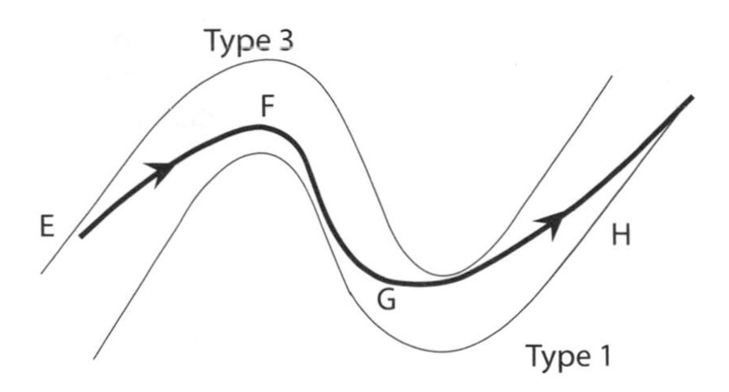

Figure 3.17 The upper figure shows our initial guess at the high-speed line through the corners when considered individually. When considered in the context of the overall track, the second turn is a type 1 turn. To maximize our exit speed, we must set up a late apex in the second turn. To accomplish this, we must sacrifice the speed and the line in the first turn.

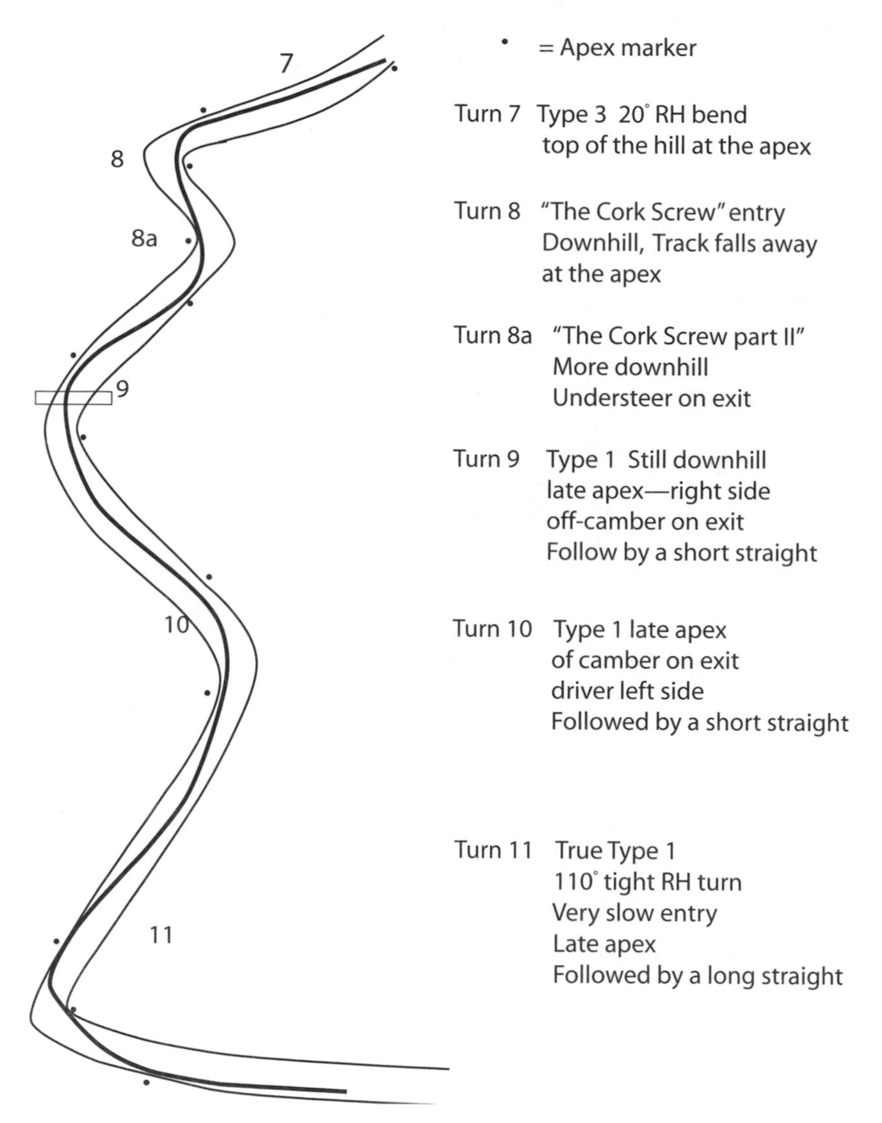

Figure 3.18 Sketch of Laguna Seca turns 7 through 11. The thick black line is the approximate racing line.

will be forced to slow. Road race circuits often use off-camber turns to increase the challenge level. Figure 3.18 is a sketch of Laguna Seca's turn 7 through 11 complex, one of America's road racing jewels. I first stood by the side of this track complex 35 years ago. It was at an SCCA race. I have watched dozens of races there, both in person and on TV. I have always wanted to drive there but have yet to get the chance. Turns 9 and 10 are both off-camber at the corner exit. They are found in the middle of a very tricky, huge downhill section. Most drivers, when faced with a new course, go looking for more information. Descriptions and video can be easily found on the Internet. In this case, the Northern California Shelby American Automobile Club has posted on the Web a nice turn-by-turn description of driving this track. It was written by Scott Griffith. The turn 7 through 11 complex starts with an uphill straight. It leads to turn 7, an uphill 20° right-hand (RH) turn. We might be tempted to treat it as a type 2 turn with an early apex and deep entry. The problem is that turn 7 marks the top of the hill and leads into a complicated downhill series of turns. While little or no time can be gained here by pressing hard, lots of time can be lost by getting out of position. Turn 7 requires a late apex to be in the proper position for entry into turns 8 and 8a. Turns 7, 8, and 8a are all type 3 turns where we compromise our speed to set up for a follow-on corner.

Turns 8 and 8a form "the corkscrew," two wicked downhill turns where the track drops away from the driver's view. The car feels light as the track drops away and the road's normal force decreases. Steering effectiveness fades and the car understeers. Get out of position or enter too fast and the car plows straight ahead and leaves the pavement. Turn 8, a left-hand (LH) turn, requires a slow entry from the RH side of the road headed to an apex on the left. The downhill angle increases at the apex. The temptation on corner exit is to drift far to the right for a fast departure. Instead, the drivers must point the car directly at the apex of turn 8a, a quick downhill RH turn. Again, the track quickly falls away at the corner apex and the car feels light. Briefly, the road flattens and the vertical load skyrockets. If the driver turns too sharply to the right during the downhill section, the car will dart right when the road flattens. As the car exits 8a and drifts left, we can squeeze the throttle and unwind the steering wheel as the downhill slope returns. Hold a bit of the RH turn of the steering wheel and cross the road to the RH side. As we straighten the wheel, short hard braking is required to prepare for the entry into turn 9. This turn suffers from a little understeer on entry. Turn 9 is followed by a short straight, making it a type 1 turn. Next is a late apex on the LH side of the track and back on the throttle followed by a drift to the right side of the road. Drift too far right and the road drops away off-camber. The off-camber has two effects. First, the normal force from the road pushes the car to the outside of the corner. More tire lateral force from the front tires will be required to keep the car on the track. Second, the off-camber road surface will increase the vertical load transfer from the inside tires to the outside tires. We will show in later chapters that load transfer to the outside wheels in a turn reduces the overall lateral force. Both effects of the off-camber road surface increase understeer.

Next, a short straight precedes turn 10, a relatively fast 90° RH bend. Again, on corner exit we drift left. Drift too far and the outside edge of the road falls away off-camber, pushing the car farther to the left and off the road. Another short straight and we must slide to the right and dive deep into the brakes. Turn 11 is a very sharp 110° LH bend. This is a true type 1 turn with a very slow entry leading to a late apex followed by a long straight. On corner exit, the car drifts to the right side of the track and narrowly misses brushing the wall. This is as good as racing gets!

Changes in the road surface, rain, curbing, the torque and power of your engine, your tires and suspension, and a change in elevation of the road can all affect the optimal line. Learning the effects of these details is one of the things that make driving the same course over and over a lot of fun.

3.9 TURNING WHILE SPEEDING UP

This is the most mathematically challenging section of this text. It is calculus based and applies mathematics beyond that used in first-semester physics. If you are not ready for this, skim the section and move on.

For a competition driver, turning while speeding up or slowing down at the limit of traction is a huge challenge. The driver must develop a feel for feathering off the brake or squeezing on the throttle while the steering wheel and the seat of his pants tells him what's happening with lateral force. Step over the limit and he begins to roast the tires and waste time or, worse, lose total

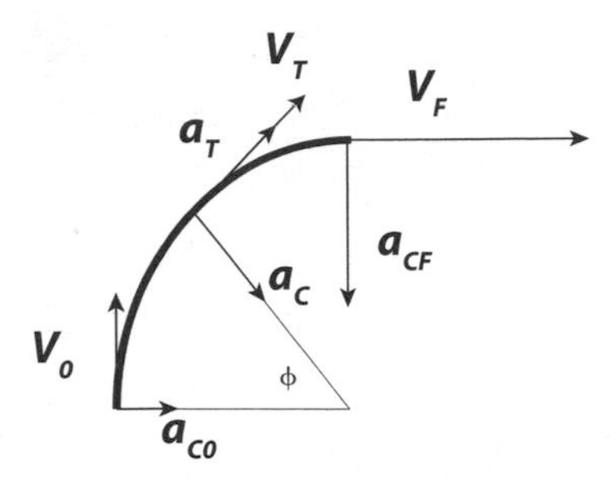

Figure 3.19 Constant-radius turn entered at speed V_0 while speeding up at a maximum rate such that the net friction force does not have to exceed $\mu_s mg$.

control. It is not a novice track driver's skill. It is, however, a champion's skill. It is worth learning a little about what happens under these conditions.

Let's consider what happens when we enter a constant-radius turn at some speed that is less than the maximum allowed by the available friction. For simplicity, we will assume that we are on a flat horizontal road. The discussion that follows here was inspired by a question that appeared in *The Physics Teacher* in 2004. The question was answered by Eugene Mosca and Scott Wiley. Carl Mungan subsequently analyzed their two results in 2006. The situation is shown in figure 3.19. A car enters the corner with velocity V_0 and attempts to speed up.

The corner radius is R and the initial centripetal acceleration is $a_{c0} = V_0^2/R$. By stepping on the gas, the driver applies a tangential acceleration, a_T. To avoid skidding, the net acceleration is limited to $\mu_s g$, as shown in equation (3.8). The square root is simply applying the Pythagorean theorem to the two components of the acceleration:

$$F_{net} = m\sqrt{a_c^2 + a_T^2}$$
$$f_s = \mu_s mg \tag{3.8}$$

or

$$\mu_s mg = \sqrt{a_c^2 + a_T^2}.$$

As the car speeds up, the centripetal acceleration increases. To avoid skidding, the driver must lift his foot slightly off the gas pedal. In the end, all of

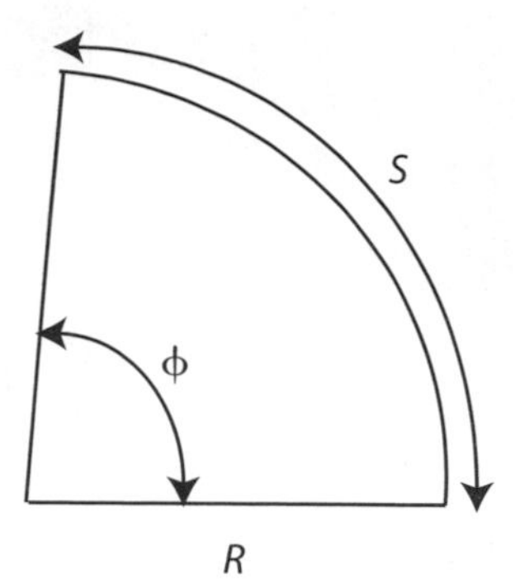

Figure 3.20 Defining the angle, ϕ, in radians in terms of the arc length, *S*, and radius, *R*.

the frictional force is committed to the centripetal component, requiring the speed to remain constant from that point onward. Given these criteria, we would like to know the tangential velocity as a function of where we are along the arc and how long it takes us to get there. We can solve for the tangential acceleration in terms of the changing tangential velocity as follows:

$$a_T = \frac{1}{R}\sqrt{V_F^4 - V_T^4} \tag{3.9}$$

and

$$a_T \equiv \frac{dV_T}{dt}. \tag{3.10}$$

It is important to describe angles in terms of radians, instead of degrees. It is worth taking a moment to refresh our memory about radians. Figure 3.20 shows a segment of a circle of arc length S and radius *R*. The arc length subtends an angle ϕ.

We define the angle in radians as the ratio of the arc length to the radius. Since both quantities have units of length, the ratio is dimensionless:

$$\phi \equiv \frac{S}{R} \text{ and in a differential form } d\phi = \frac{1}{R}dS.$$

We can relate radians to degrees by considering a circle. The arc length for an entire circle (360°) is the circumference, which is equal to $2\pi R$. The ratio is $2\pi R/R$, meaning that 2π radians are equal to 360°. Similarly 180° is π radians and so forth.

The next few steps require a little calculus to complete the calculation. If you are not experienced with calculus, jump to the result in equation (3.16). If we use the arc length to define the position of a car, then the rate of change of the arc length is the tangential velocity:

$$V_T = \frac{dS}{dt}, \tag{3.11}$$

which we can rearrange as

$$dt = \frac{dS}{V_T}. \tag{3.12}$$

Combining our two equations for a_T with this last result and the differential form of the angle, we have

$$\frac{V_T}{R}\frac{dV_T}{d\phi} = \frac{1}{R}\sqrt{V_F^4 - V_T^4}. \tag{3.13}$$

The R drops out and we can perform a separation of variables on the remainder. With ϕ on one side and V_T on the other, we can integrate both sides. The primes indicate dummy variables of integration:

$$\int_0^{\varphi} d\phi' = \int_{V_0}^{V_T} \frac{V_T' \, dV_T'}{\sqrt{V_F^4 - V_F^{4\,\prime}}} \tag{3.14}$$

and

$$\phi = \frac{1}{2}\sin^{-1}\left[\frac{V_T^2}{V_F^2}\right] - \frac{1}{2}\sin^{-1}\left[\frac{V_0^2}{V_F^2}\right]. \tag{3.15}$$

The second term is a constant that depends on the entry speed as well as the maximum speed without skidding. The angle is a maximum when the tangential speed is equal to the maximum speed. The inverse sine of 1 is $\pi/2$ (equivalent to 90°). This yields the angle required to achieve the maximum speed:

$$\phi_{\max} = \frac{\pi}{4} - \frac{1}{2}\sin^{-1}\left[\frac{V_0^2}{V_F^2}\right]. \tag{3.16}$$

This is an interesting result. *If the car starts from rest, $V_0 = 0$, the car will cover one-eighth of the circle ($\pi/4$ radians) before reaching maximum speed, independent of the radius of the circle or the coefficient of friction.* The greater the entry speed, V_0, the smaller the fraction of the circle it can cover. The importance of opening up the exit radius on a type 1 turn becomes clearer. A larger radius means a larger arc length for a given angle. This means that we can accelerate over a larger distance and exit with a greater speed. We can rearrange equation (3.15) to obtain an expression of the tangential speed as a function of angle:

$$V_T = V_F \sin^{1/2}(2\phi + C), \tag{3.17}$$

where C is a constant

$$C = \frac{1}{2}\sin^{-1}\left[\frac{V_0^2}{V_F^2}\right]. \tag{3.18}$$

You will recall that

$$V_T = \frac{dS}{dt} = R\frac{d\phi}{dt}. \tag{3.19}$$

We can combine this with equation (3.17), separate variables, and integrate again to obtain the amount of time over which we will be able to accelerate:

$$t_f = \frac{R}{V_F}\left[\sin^{1/2}(2\phi + C) - \sin^{1/2}(C)\right]. \tag{3.20}$$

If we use our example type 1 turn from figure 3.13, we can fill in some of the values:

$$V_0 = \sqrt{\mu r_{en} g} \tag{3.21}$$

$$V_F = \sqrt{\mu R_{ex} g} \tag{3.22}$$

$$C = \frac{1}{2}\sin^{-1}\left(\frac{r_{en}}{R_{ex}}\right). \tag{3.23}$$

What happens if we increase our exit radius by a factor of 2? We increase our exit speed by a factor of 1.4. The max angle, ϕ_{max}, through which we can accelerate without skidding is 30°. For a 50 m radius and a static coefficient of friction of 1.2, we have about 2 s of controlled, tapering acceleration.

3.10 SUMMARY

We started the chapter with goals focused on understanding the proper racing line. We learned the importance of vector analysis and how to take advantage of the width of the track. We learned a method for classifying turns based on their relative importance. We explored the driving and mathematical challenges of simultaneous turning and tangential acceleration.

Lessons learned:

- The traction circle and the g-g diagram are valuable tools in analyzing car and driver performance.
- We analyzed an isolated corner and learned the value of using the road to open up the turn radius. We used the isolated corner model to construct a simple track.
- The isolated corner model track was given variable straightaway lengths and variable car acceleration abilities. This led us to the idea of low-power "momentum" cars.
- A type 1 turn at the start of a straight can be optimized with a late apex and opening the radius of the turn on exit. This is only effective if you can increase the exit speed. This type of turn takes more time. It is only valuable if the increased exit speed leads to a long follow-on straight.
- A type 2 turn at the end of a straight uses an early apex and deep entry. Maintain speed for as long as possible.
- A type 3 turn is a transition turn. It is easy to lose time here and hard to gain time. The goal here is to set up the type 1 turn that will eventually follow.
- Our strategy must be modified by other considerations, such as minimizing the impact of off-camber turns.

Chapter 4

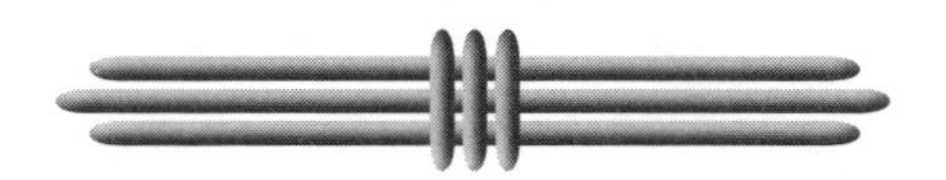

Basic Vehicle Dynamics: Load Transfer and Tires

4.1 CENTER OF GRAVITY

The most fundamental physics quantity involved in vehicle dynamics is the center of gravity. In physics, we tend to talk about the center of mass, while engineers talk about the center of gravity (CG). The good news is that as long as the local gravitational field is uniform, the two are identical. Like everyone else, we'll use the terms interchangeably. The center of mass is the mass-weighted average position of the mass of a system. As far as gravity is concerned, we can treat the system as if all of the mass were located at this single point. For a string of small masses strung out in a line, we add up the product of the particle mass, m_i, times the *x*-coordinate, x_i, for its position and divide by the total mass, M_T:

$$x_{\text{CM}} \equiv \frac{\sum m_i x_i}{M_T}.$$

Of course, a car is a three-dimensional object, so we will need to add to this a calculation to the *y* and *z* centers of mass:

$$y_{CM} \equiv \frac{\sum m_i y_i}{M_T} \qquad z_{CM} \equiv \frac{\sum m_i z_i}{M_T}.$$

For the parts that aren't small, we will need to do an integral. Since most of us are going to use production cars and not design them, both the sum and the integration seem a bit challenging. Instead, we will measure a few parameters for the car and do an equilibrium calculation to find the center of gravity. We will do three separate calculations to find the longitudinal CG, the lateral CG, and the height of CG. Let's assume that all the fluids are full and the tires are at operating pressure.

4.2 LONGITUDINAL AND LATERAL CENTER OF GRAVITY

Figure 4.1 is a side-view free-body diagram for our car at rest. The dimension labeled L is the wheelbase of the car and our unknown quantity is d, the horizontal distance from the front wheel hub center to the center of gravity. Five

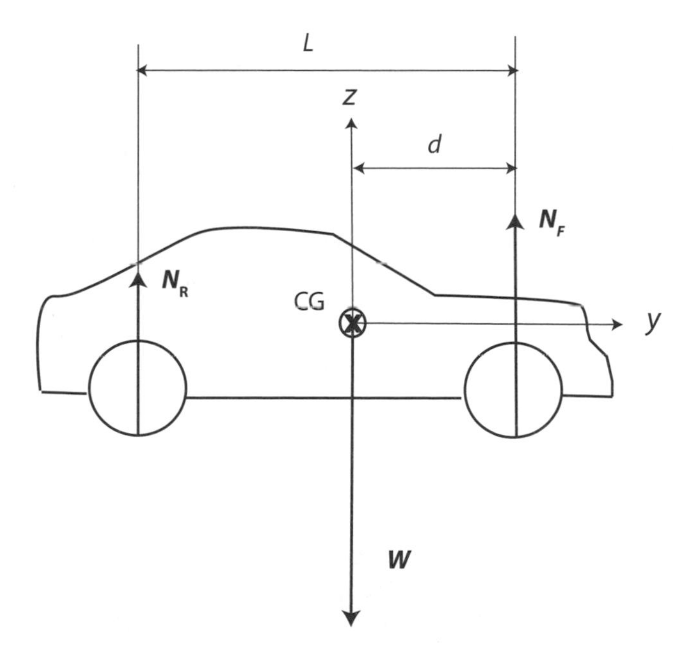

Figure 4.1 Side-view free-body diagram for a car at rest. N_F and N_R are the front and rear normal forces at the tires, respectively. W is the weight. L is the wheelbase of the car, and d is the unknown location of the CG behind the midpoint of the front wheel.

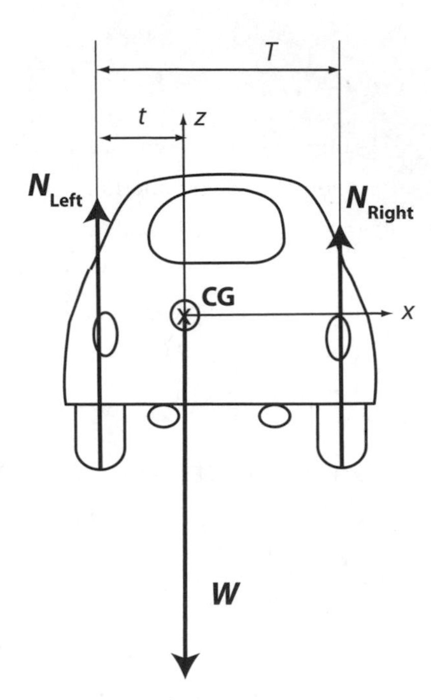

Figure 4.2 Rear view of the static forces acting on the car. *T* is the wheel track and *t* is the distance from the left wheel center to the center of gravity.

external forces act on the car, gravity and a normal force at each tire. For the purpose of finding the distance *d*, we will combine the two front normal forces as well as the two rear normal forces. For the equilibrium calculation, we will sum the forces by component and set them equal to zero, as well as summing the torques and setting those equal to zero. Since the forces act only in the *z*-direction, two of the sums are eliminated:

$$\begin{aligned}
&\sum F_z = 0 \\
&N_F + N_R - W = 0 \\
&\sum \tau = 0 \\
&N_F d - N_R(L - d) = 0 \\
&d = L\frac{N_R}{W}.
\end{aligned} \tag{4.1}$$

How do we determine values for the forces to solve this equation? We measure the four normal forces by placing a scale under each wheel on a level floor.

The sum of the normal forces is equal to the weight. The distance d will then help us to evaluate the load transfer under braking.

Figure 4.2 is the back view of the car. The quantity T is called the wheel track or simply the track. N_{Left} is the sum of the normal forces on the two left-hand tires, and N_{Right} is for the right side. We can apply an identical procedure to the longitudinal case for this problem and solve for t, the distance to the centerline of the left-hand wheels. For a symmetric car $t = T/2$. A driver is probably the largest asymmetry in most cars. The battery is next in mass. It still should be nearly $T/2$. The resulting expression for t is

$$t = T\left(\frac{N_{\text{Left}}}{W}\right). \tag{4.2}$$

4.3 HEIGHT OF THE CENTER OF GRAVITY

To find the height of the center of gravity, we will again perform a static equilibrium calculation in the longitudinal direction. The scales won't help us find the height unless we elevate the front wheels and compare the scale readings with the horizontal case. In figures 4.3 and 4.4 we have significantly simplified the sketch of the car to allow us to focus on the geometry.

We need to address several points of complication when elevating the front end. It is tempting to set up the static equilibrium problem such that the height of the CG is referenced with respect to the ground. After all, it is the height above the ground that we need to know for dynamic analysis. However, as we raise the front end with the rear brakes locked, the contact points between the tires and the ground move counterclockwise around the tires, complicating the contribution of the tires and wheels. The good news is that the line of action for the normal forces still passes through the center of the wheel and hub. If we perform the static equilibrium calculation referenced to the center of the wheel, we can simply add the radius of the wheel after we are finished. The second problem is that the springs at the rear compress when the front end is elevated, changing the geometry of the car. Race teams avoid this problem by replacing the spring and shock assembly with a solid link while performing this test.

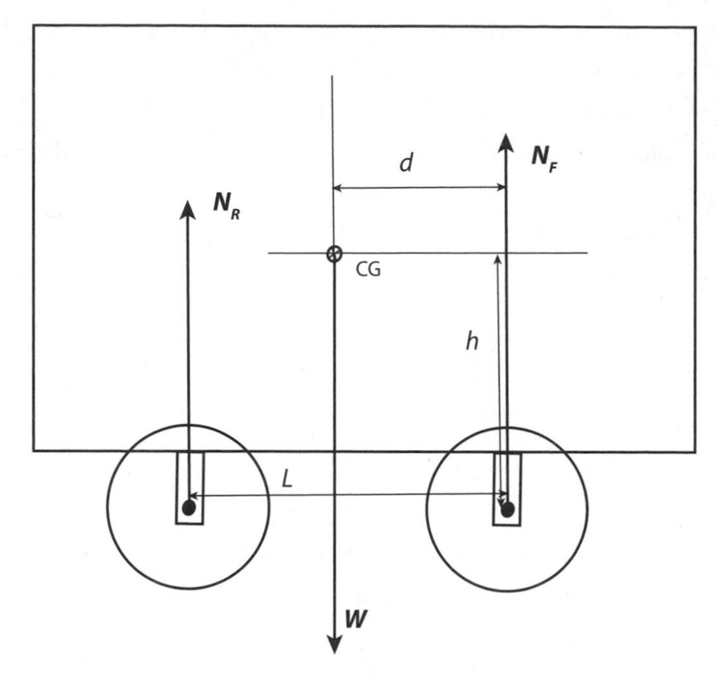

Figure 4.3 The height of the center of gravity, *h*, can be calculated from the normal force distribution in this orientation if it is combined with the results of the case in which the front wheels are elevated. The distance *h* in this calculation is with respect to the wheel centers.

For the inclined figure, figure 4.4, there are essentially no horizontal forces. The vertical forces in equilibrium tell us, once again, that the weight is equal to the sum of the normal forces. Summing the torques about the center of gravity, we have

$$BN_{F1} - AN_{R1} = 0, \tag{4.3}$$

where A and *B* are the distances from figure 4.4. Subscript 1 indicates that the front and rear normal forces are from the inclined case. A zero subscript will indicate the normal forces from the horizontal orientation. The next several relationships come from inspecting the inclined figure and considering the variety of right triangles that exist:

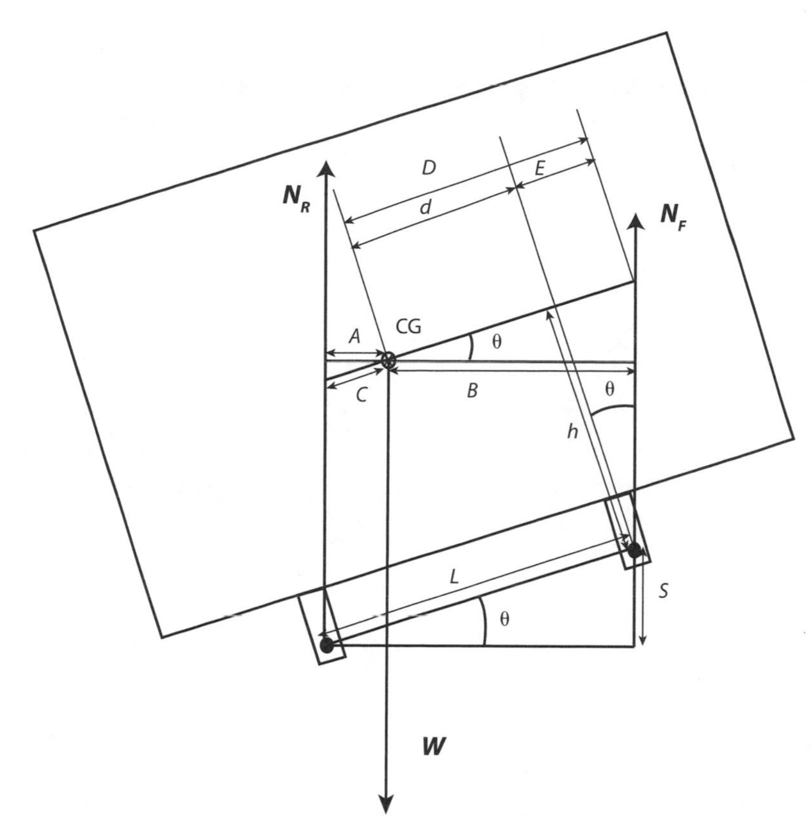

Figure 4.4 This is our simplified car with the front end elevated a distance *S*. The wheels are removed from the figure for simplicity.

(i)	$\cos(\theta) = B / D$	
(ii)	$\cos(\theta) = A / C$	
(iii)	$C + D = L$	
(iv)	$E + d = D$	
(v)	$B = D\cos(\theta)$	From i
(vi)	$B = (E + d)\cos(\theta)$	Combine iv and v
(vii)	$\tan(\theta) = E / h$	
(viii)	$B = (h \tan(\theta) + d)\cos(\theta)$	Combine vi and vii
(ix)	$\dfrac{A}{\cos(\theta)} + \dfrac{B}{\cos(\theta)} = L$	Combine i, ii, and iii
(x)	$A = L\cos(\theta) - B$	From ix

(xi)	$BN_{F1} - (L\cos(\theta) - B)N_{R1} = 0$	From eq. (4.3) and x
(xii)	$B(N_{F1} + N_{R1}) = L\cos(\theta)N_{R1}$	Rearrange xi
(xiii)	$BW = L\cos(\theta)N_{R1}$	$\sum$ normal force = weight
(xiv)	$(h\tan(\theta) + d)\cos(\theta)W = L\cos(\theta)N_{R1}$	combine xiii and viii
(xv)	$(h\tan(\theta) + \frac{LN_{R0}}{W})W = LN_{R1}$	
(xvi)	$(h\tan(\theta)W = L(N_{R1} - N_{R0})$	
(xvii)	$h = \frac{L\Delta N_R}{W\tan(\theta)}$	(4.4)

While this is a somewhat painful exercise in trigonometry and algebra, it proves the equation for the most useful center of gravity parameter, the height of the center of gravity. This exercise has bogged down many a student along the way.

4.4 LOAD TRANSFER AND THE STATIC STABILITY FACTOR

The study of physics shares language with our everyday vocabulary. For example, in physics we define "work" as a force, acting over a distance. In our daily life, we roughly define work as doing something constructive (and boring if you believe my son). Work can even be a mental activity in our general language, which is a far cry from the physics usage. The two definitions can create confusion. Another such term is "weight transfer." In physics, weight is the force of gravity acting on an object. It is equal to the product of the mass times the acceleration of gravity. The only way to "transfer weight" is to actually add or subtract mass from the car. Or perhaps you might transfer mass from one end of the car to the other. However, this isn't what racers mean when they talk about weight transfer. They mean a redistribution of the normal force between the tires due to an acceleration of the car. To be consistent with our definition of weight, we are going to avoid this term "weight transfer" and instead use the phrase "load transfer" or "vertical load transfer."

In our discussion of the friction circle in chapter 2 we introduced the idea of an inertial term or an inertial force. This is a fictitious force that appears to an observer moving in an accelerated reference frame. The load transfer case that we are going to consider first is a car under braking. When the driver going 60 mph slams on the brakes, the passengers feel like they are being thrown forward by a force. If we change reference frames to one fixed at the side of the road and not accelerating with the car, we immediately see what is happening. The car is slowing but the body of the passenger is still trying to go 60 mph. His body has inertia, and according to Newton's first law, it will continue at 60 mph until acted upon by some external force. For the passenger that force comes from the seat belts and whatever else the passenger is in contact with inside the car. The nonaccelerating inertial frame of reference allows us to see that the true force on the passenger is the car pushing backward on the passenger, in an effort to slow him down. This same problem arises in circular motion. A driver turns sharply to the left at 60 mph. A passenger in the same car "feels like" he is being pushed to the outside and slammed against the car's right-hand door by some force. If we step out of the reference frame of the car, we see what is really happening. The passenger's body is, according to Newton's first law, trying to continue in a straight line, tangent to the circle. It is the car door that is pushing the passenger inward, pushing his body into a circular path.

The passenger never goes anywhere with respect to the car. Being a physics student, this passenger wants to apply static equilibrium conditions in the reference frame, summing the forces and setting them equal to zero. The only way to make this work is to add a made-up force to the equation that is equal and opposite to the acceleration of the reference frame:

$$\sum \boldsymbol{F} - m\mathbf{A} = m\boldsymbol{a}.$$

The vector $\boldsymbol{a}$ is the observed acceleration in the car's frame and $\mathbf{A}$ is the acceleration of the reference frame itself. Let's consider what happens under braking. Figure 4.5 is our side view of the car with the addition of friction between the road and the tire and the inertial force, $m\mathbf{A}$. The friction force points opposite to the direction of motion. The passengers experience a forward pointing inertial force.

Once we know the location of the center of gravity, we can calculate the load transfer. Combining the static equilibrium equations yields an expression for the normal force at the front wheels:

$$\sum F_y = -f_F - f_R + mA_y = 0$$
$$\sum F_z = N_R + N_F - mg = 0$$
$$\sum \tau_{CG} = N_F d - N_R(L - d) - (f_F + f_R)h = 0$$
$$N_F = mg\left[1 - \frac{d}{L}\right] + mA_x\left[\frac{h}{L}\right]. \tag{4.5}$$

The first term is the normal force before braking. The ideal distribution would be to start with a 50/50 distribution of load between the front and rear wheels at rest. In a front engine car *d* is biased toward the front wheels, making the first term greater than half the weight of the car. The second term is the additional normal force on the front wheels due to the braking. On street tires, the braking acceleration is limited to about 1 g. This makes *h*/*L* a critical parameter in determining brake bias. For example, a 24 inch height of the center of gravity for a 120 inch wheelbase adds to the normal force an amount

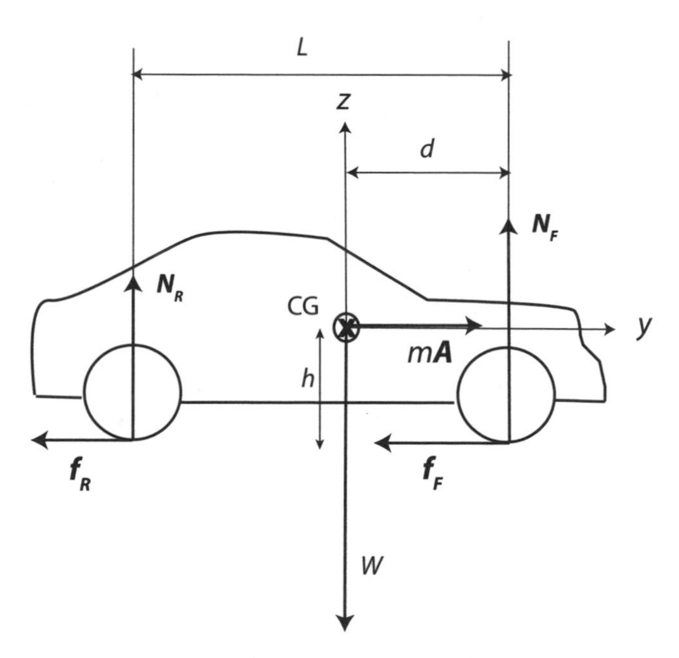

Figure 4.5 Free-body diagram for a car under braking in the non-inertial reference frame of the car. Note the addition of the inertial force *mA*.

equivalent to 20% of the car's weight to the front tires. For the ideal 50/50 static load distribution, this would place 70% of the vertical load on the front tires and 30% on the rear. This is the reason the nose of the car dips under braking when the increased front load compresses the front springs. Since normal force is proportional to friction force, the front wheels must accomplish 70% of the braking. To achieve balanced braking and avoid rear-wheel lockup, designers must proportion the braking force between the front and rear in accordance with the normal force load transfer. Designers frequently employ smaller brake disks on the back wheels, as well as sending a lower brake system hydraulic pressure to the rear calipers.

Load transfer happens when a car turns as well. Consider a left turn with the addition of frictional forces pointing to the left at each wheel and the resulting inertial force pointing to the right. Load transfers to the right wheel with the following normal force:

$$N_R = W\frac{t}{T} + mA\frac{h}{T}. \tag{4.6}$$

The first term is the static load on the right wheel. The second term is the resulting load transfer due to turning. This effect is more significant than the longitudinal case because the track width, T, is much less than the wheel base, L. Consider the special case where the entire load is transferred to the right wheels and $N_R = W$. For simplicity let the lateral center of gravity be in the center of the car and $t = T/2$. Solving for the required acceleration,

$$A = g\left[\frac{T}{2h}\right].$$

This is the rollover threshold. Lateral accelerations greater than this quantity will lead to a rollover accident. The National Highway Transportation Safety Administration (NHTSA) refers to the quantity in brackets ($T/2h$) as the Static Stability Factor (SSF) and quotes values for passenger cars of 1.3 to 1.5. This is an acceleration of 1.3 to 1.5 g's, which is much greater than the typical tire can deliver. This means that a car will slide or skid rather than roll over. Sport utility vehicles (SUVs) are taller and slightly wider than a passenger car. The SSF for SUVs is quoted as being in the range of 1.0 to 1.3. They are still unlikely to roll over because of skidding. The NHTSA used the SSF as the sole input to their rollover resistance rating until 2004. They have since added a dynamic test to the procedure. Most rollover accidents involve a tripping incident where the skidding tire encounters an obstacle such as a curb that stops the skid. But a low SSF remains an indicator of a higher probability for rollover.

We have ignored the fact that the load transfer compresses the springs and tires on the side of the car where the normal force has increased. At the same time, the inside springs and tires expand. The resulting body roll introduces further complications, which we will address as we explore the function of the suspension and tires.

4.5 TIRES AND FORCES

Tires are the limiting factor in automotive performance. They support the weight of the car, accommodate road irregularities, thrust the car forward under acceleration, claw it to a stop under braking, and change its direction under steering. When they fail to do their job, the consequences are staggering. In June of 2005 at the Indianapolis Motor Speedway, with 110,000 people sitting in the stands and 20 million watching on TV, 14 of 20 Formula 1 cars pulled into pit lane and failed to start the race. The reason? Tire flaws. Investigations following crashes that had occurred during practice convinced the tire manufacturer to declare their tires unsafe. With no suitable substitutes on hand, the racing team managers ran out of options and ordered their drivers into the pits. Another major incident was highlighted in May of 2001 when the Department of Transportation's NHTSA opened an investigation into tire failures on SUVs. Following the death of 271 people in SUV rollover accidents, NHTSA directed the recall of 6.5 million tires, the first of many such recalls. Tire stories are not hard to come by.

Physics students face an interesting dichotomy of complexity and simplicity. Unwinding the intricate fabric of mathematics and physics principles is a demanding task. At the same time, physics practice problems are the epitome of oversimplification. The study of systems of frictionless blocks pulled by massless inextensible ropes passing over the massless, frictionless pulleys has, at best, limited applications. But this kind of simplification is at the heart of our method. We stick with simplified models until they fail to explain system behavior. That's when the physics problems get challenging, and that's where we are with tires. They aren't massless. They aren't rigid. They flex, twist, bulge, and stretch in response to applied forces. They are about as far from homogeneous as you can get. Part rubber, part metal, part fabric, and part air, tires are possibly the most complex factor affecting a car's performance. However, beneath this shell of complexity lie the rules of fundamental physics.

4.6 TIRE CONSTRUCTION

Figure 4.6 shows the cross section of an automobile tire and includes the general features needed to begin to understand the physics. The bead is a hoop of metal wire found in the inner diameter of the tire. The bead is coated in rubber and holds the tire tight against the metal wheel. The body plies are sheets of woven cords contained within the rubber to add strength to the tire. The cords are made from everything from polymers to natural textiles. For radial tires, the body cords run from bead to bead, in the clockwise direction in figure 4.7. When viewed from the side of the tire, they appear to run in the radial direction (from bead to tread).

The inner liner forms the airtight inner boundary. The rubber surrounding the bead is pressed against the metal wheel and forms a pressure seal. The metal of the wheel completes the pressurized boundary. The tread forms the outer circumference of the tire. The grooves, blocks, ribs, and sipes that make up the tread are designed to remove water from beneath the tire and allow the rubber to remain in direct contact with the road. This improves traction and, in the extreme case, prevents hydroplaning, where

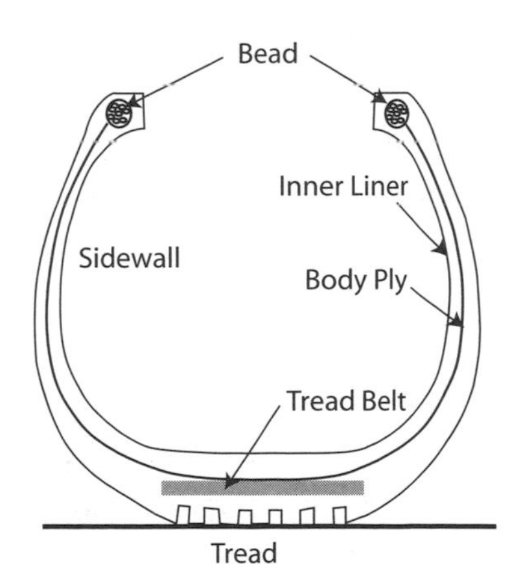

Figure 4.6 Tire cross section details.

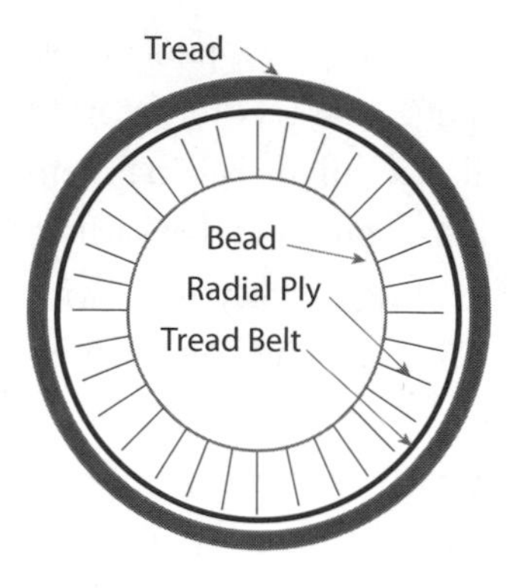

Figure 4.7 Tire construction sidewall details.

the tire rides up on a cushion of water, resulting in a catastrophic loss of traction. The tread belts found beneath the treads are designed to shape the tread and to provide strength and stability to the tread. Tread belts are made from everything from fabric to steel. An unloaded tire cross section without tread belts would be almost semicircular. The addition of tread belts flattens the outer surface of the tire from left to right in figure 4.6.

The rubber that forms the surface of the tire is an elegant product of chemical engineering. It must be flexible, durable, thermally and mechanically stable, adhesive, elastic, and chemically stable when exposed to the environment.

4.7 WHEELS

The tires are mounted to the car via the wheels, which are typically made of steel or aluminum or a variety of other lightweight alloys. Figures 4.8 and 4.9 show a few basic dimensions that we need to understand if we ever want to replace the original tires and wheels. The bolt circle diameter is measured through the center of the bolt holes of the four- or five-lug pattern. The hole for the hub helps to keep the wheel centered. The wheel width and diameter are measured at the point where the tire bead seats in the wheel. The last quantity of interest is the wheel offset. This is the distance from the wheel centerline to the inside mating surface for the hub shown in figure 4.8. This is a critical parameter. It determines how far the wheel is set into or out of the wheel well of the car. A zero offset puts the mating surface at the wheel centerline. A positive offset moves the mating surface toward the roadside, whereas a negative offset moves the mating surface away from the roadside and deeper into the wheel well. It therefore determines how much clearance the tire has to the fender and suspension components. It also determines the scrub radius, which we will go over in the suspension section. Scrub radius has a large effect on the steering characteristics of the car. Not all wheel manufacturers use offset as a design parameter. Some use the parameter "backspace." Backspace is the

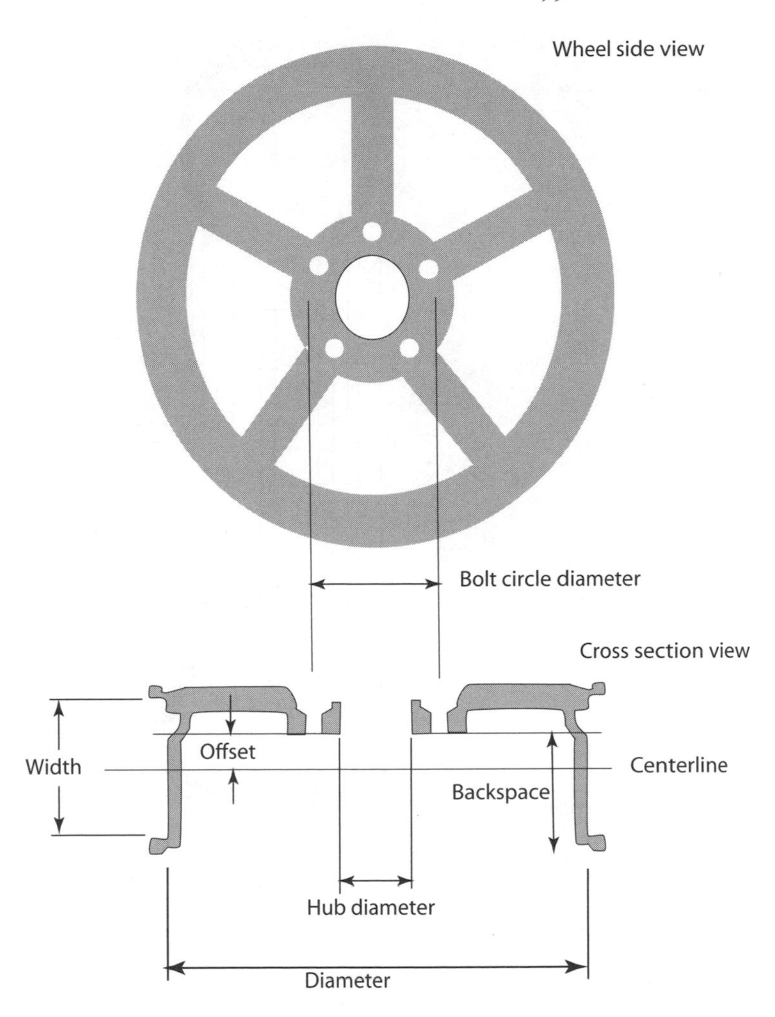

Figure 4.8 Wheel construction details.

distance from the hub mating surface to the inside plane of the wheel. They use it because it is easy to measure.

4.8 TIRES UNDER STATIC LOAD

Figure 4.9 shows the cross section of a tire and wheel combination under static load. What can we learn about the function of the tire from this figure?

Consider a car at rest on a flat horizontal surface. Earlier in this chapter

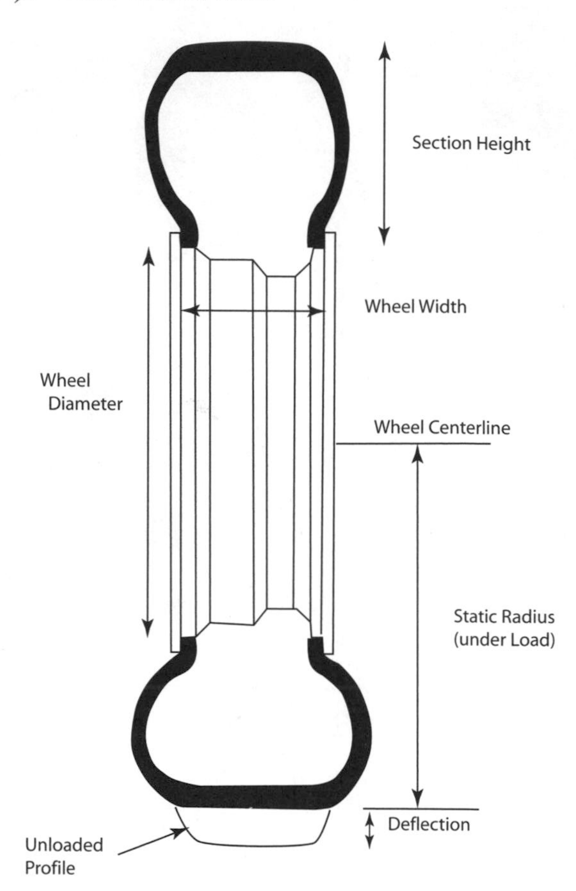

Figure 4.9 Cross section of a loaded tire.

we found that five external forces act on the car at rest. Gravity pulls vertically downward on the car, and at each tire, the ground exerts a vertically upward normal force, **N**. (In mathematics, normal means perpendicular to the surface. Since the car is at rest and the ground is flat, no tangential forces act at the ground-tire interface. We'll consider tangential forces later.) Because the mass is distributed unevenly about the car, the four normal forces are rarely identical, although left-right symmetry is nearly achievable. Engine location dominates the front-to-back load distribution. From Newton's second law we have shown that

$$\sum F_z = ma_z$$

$$N_{LF} + N_{RF} + N_{LR} + N_{RR} - mg = 0.$$

An unloaded tire viewed from the side forms an almost perfect circle. When the wheel and tire are mounted on a car and placed under load, the surface of the tire in contact with the ground flattens as shown in figure 4.9. The flattened surface is referred to as the contact patch. The normal force of the ground acting on the tire is distributed over the contact patch. While the distribution of the normal force over the contact patch is not perfectly uniform, it is still a reasonable first approximation. The force per area is defined as a pressure with units of N/m^2 or lbs/in^2 (psi). For example, a typical passenger car inflation pressure is 35 psi. More properly this should be referred to as a gauge pressure (pounds per square inch gauge, psig) since it is a pressure above atmospheric pressure and not an absolute pressure (pounds per square inch absolute, psia). Our second approximation is that the pressure exerted by the normal force is approximately equal to the inflation pressure of the tire. We can show this by considering a square inch of tire surface that makes up the contact patch. As long as we are not on the edge of the contact patch, the only vertical forces are the pressure downward and the normal force upward (neglecting the weight of the contact patch rubber). If we know the weight of our car, we can estimate the size of the contact patch.

Let's consider an example:

$$P = \frac{F}{A} \qquad A = \frac{F}{P} = \frac{3500 \text{ lbs}}{35 \text{ psig}} = 100 \text{ in}^2.$$

Hence, 100 in^2 distributed over four tires, or 25 in^2 per tire, is a reasonable estimate of the size of the contact patch. It is important to keep in mind that this is an approximation. The normal force across the contact patch is not completely uniform. For an example, an underinflated tire tends to pucker upward at the center of the contact patch and reduce the normal force in this region. If allowed to continue in the underinflated state, the tire will prematurely wear on the outside edges. Overinflation will produce the opposite result.

Racers use a tire pyrometer to measure temperatures across the face of the tire to determine the state of loading across the contact patch. A metal needle in contact with a thermocouple is stabbed into the tire tread and into contact with the tread belts. A constant temperature in three different places across the width of the tread is an indicator of even loading. Some racers will use an infrared thermometer to measure tire surface temperatures. They do this because it is quick, easy, and relatively inexpensive. The problem with this technique is that the tire surfaces cool quickly. For safety reasons autocrossers typically have to drive at a walking pace from the finish line back to the grid. It takes a minute or two to get to the tires. Road racers usually take a cooldown lap to allow the brakes and drivetrain a chance to cool uniformly before stopping. The tires cool as well. The tire core is more likely to retain heat and give an accurate and repeatable temperature profile of the tire in use. Road racers heat the tire to the core. Autocrossers skid and slide the tires a great deal, and their laps are short, a minute or less. The tire core temperature is likely to be 100°F cooler than a road racing tire core. On the other hand, the tire surface in autocross can easily overheat and become greasy. If you can get to the tires quickly after an autocross run, a surface infrared temperature can be a useful indicator.

Interestingly, adding static load to a tire will not significantly change the tire pressure. Adding load causes the top of the tire to bulge as the bottom flattens, keeping the tire volume approximately constant. From the ideal gas law ($PV = nRT$), the pressure must be approximately constant if the volume and temperature are constant. Check your tire pressure and then jack up the front end until the tire is off the ground. Measure the pressure again. It is common to observe a pressure change of approximately 0.2 psi out of 35 psi, a change of less than 1%. The contact patch, on the other hand, changes in proportion to the load. This behavior is quite different from the rigid solids studied in introductory physics and should encourage us to be cautious in how we apply our prior knowledge to this new system.

A change in temperature of the tire will produce drastic changes in the tire pressure. People who autocross execute one lap at a time, and tire pressures rise with each lap. Drivers must vent air from the tires between each lap to maintain a constant racing tire pressure. Road racers must start with underinflated tires and wait for pressure to rise and for handling to improve.

Cars are supported via the normal force acting on the tire contact patch, but how does the tire apply forces to the rest of the car? The tire acts on the metal wheel, which is bolted to the car. It is tempting to jump immediately to the conclusion that it is the air pressure inside the tire that supports the wheel and leave it at that. This is only indirectly correct. To understand the working of the tire, we must understand the function of the air pressure. The air within the tire completely surrounds the metal wheel. Any force generated by the air pressure that acts upward on the bottom metal surface of the wheel is canceled by an equal and opposite pressure force acting downward on the upper surface of the wheel. The rubber tire must support the wheel, but how? Over the next few pages we will try to show that the tension in the side wall is a critical parameter and that the tension transmits forces to the car via the tire bead ring. We will show that the inflation pressure pretensions the sidewall, allowing it to support the weight of the car from above, and that the tire contact patch transmits forces to the wheel via the tension in the sidewalls. Let's give it a try.

You will often hear that the sidewalls support the tire. The sidewalls of an uninflated tire are soft and compliant. You can collapse them by hand. It is the inflation of the tire that adds strength and rigidity to the sidewall. Consider figure 4.10, which shows an inflated ball. Every point on the surface of the ball is under tension. The eight tension vectors shown in the figure represent the tension force, pulling tangent to the surface at a single point. In fact, we could draw a similar sketch for any point on the surface. Given that the tension is nearly uniform in all directions tangent to the surface, a more accurate picture would be a cone of tension vectors.

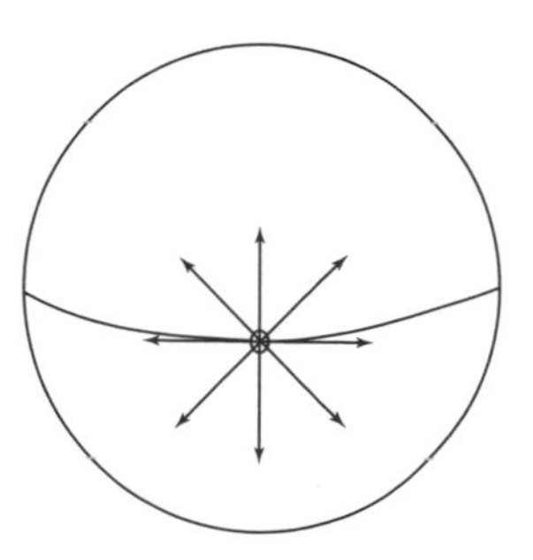

Figure 4.10 Surface tension for an inflated ball.

Let's consider a basketball at rest on the ground. Figure 4.11 shows the free-body diagram for a small element of mass on the side of the ball. $\boldsymbol{P_A A}$ is the force vector that results from the internal absolute pressure of the ball acting on the cross-sectional area of the element of mass. Atmospheric pressure exerts an opposite force, $\boldsymbol{P_0 A}$, from outside of the ball. In response to the net pressure, a cone of tension develops in the elastic surface of the ball. Two of the infinite array of tensions that make up the cone are explicitly shown in the figure (T_1 and T_2). The weight of the mass element, mg, pulls downward. Applying Newton's second law for equilibrium conditions, the vector sum for these forces must equal zero. The pressure contribution to the horizontal force

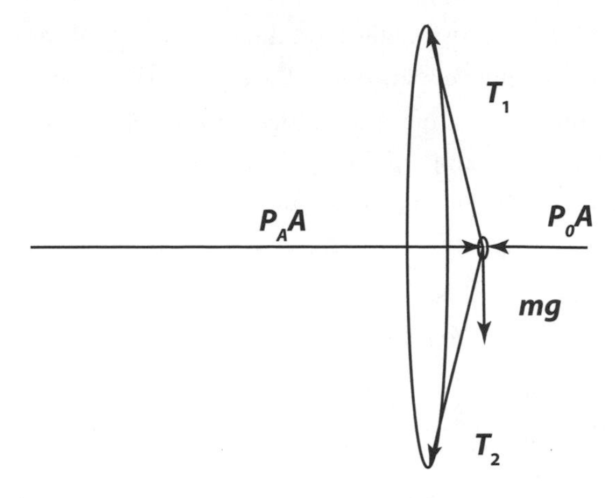

Figure 4.11 Free-body diagram for an element of mass, *m*, and area, *A*, located at the equator of the ball. The cone of tension vectors in the elastic surface develops in response to the internal pressure of the ball.

is cancelled by the sum of the horizontal components of the tension. What cancels the weight? The upward-directed half of the cone of tension must be slightly greater than the downward half. The cone is not quite symmetric.

Based on this free-body diagram in figure 4.11, the element of mass is supported by the tension from above and not by a force from below acting upward. How can it be supported from above? Figure 4.12 shows the force acting on the upper hemisphere of the ball.

If you have some experience with multiple integrals in calculus, calculating the net vertical force created by the pressure on the upper hemisphere is relatively straightforward in spherical coordinates:

$$F_z = \int P_z dA = \int_{\theta=0}^{\pi/2}\int_{\phi=0}^{2\pi}\left(P_g \cos(\theta)\right)\left(R^2 \sin(\theta)\right) d\phi d\theta = P_g \pi R^2.$$

Thankfully, the result can be applied without the use of calculus. The product of the gauge pressure times the cross-sectional area of the ball is the total upward force on the hemisphere. Applying Newton's second law, we have

$$\Sigma F_z = \frac{m}{2} a_z$$

$$P_g \pi R^2 - \frac{mg}{2} - T = \frac{ma_z}{2}.$$

The pressure acts upward and gravity acts downward. By symmetry, all the tension forces about the edge must have equal magnitudes. Since the lower edge of the hemisphere is the equator, all the tension vectors act straight down. The *T* in the equation is the sum of all the tensions around the equator. Clearly, the net internal gas pressure supports the top half of the ball. If the ball is in free fall, the tension is equal to pressure force. If the ball is at rest on a surface, the tension is reduced by the weight acting on the sphere. At the same time the pressure will increase as a result of lost volume when the ball flattens a little against the surface. Which contribution dominates? If the radius and temperature remain constant, it can be shown that the tension decreases slightly.

Let's return to our discussion of the tire and wheel. What supports the wheel? The wheel is supported by the two bead wire loops in the tire. The tension in the sidewall controls the motion of the bead. We need to understand tension in the sidewall at least qualitatively. Figure 4.13 shows a cross section for the tire below the wheel, both loaded and unloaded, as well as a cross section of the tire above the wheel when the tire is under load. By unloaded we mean that the corner of the car has been jacked up and the tire is no longer in contact with the road. By loaded we mean that we remove the jack so that that corner of the car is experiencing its full normal force in static conditions.

The weight of the car causes the wheel to exert a downward force on the tire. The lower sidewalls of the tire experience a compression that decreases the tension in the side. The internal pressure increases only slightly. The inter-

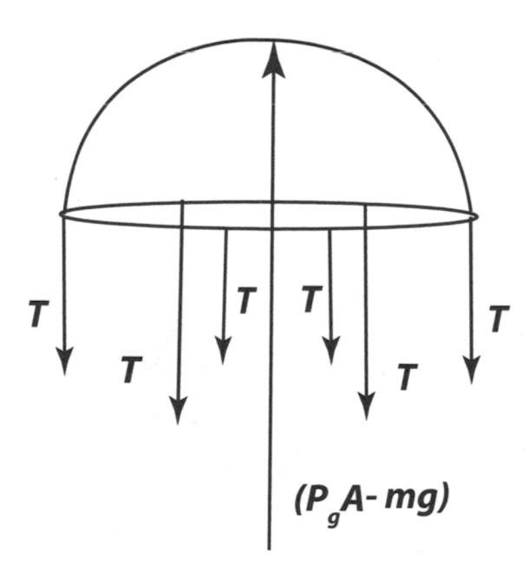

Figure 4.12 Forces on the upper hemisphere of a ball.

nal pressure overcomes the weakened tension and the tire bulges outward as shown in figure 4.13. The bulge rotates the tension vectors toward the horizontal, increasing their component in opposition to the pressure (compare the vectors in the unloaded and loaded cross sections in fig. 4.13). The bulge pushes outward until vector equilibrium is restored. At the bottom of the metal wheel the increased load pushes downward on the tire bead wire loop. In turn, the bead loop pulls downward on the upper tire, increasing the tension in the upper sidewall (see the upper tire loaded cross section vector diagram in fig. 4.13). The sidewall flattens and rotates the tension vector toward the vertical, reducing the component in opposition to pressure force.

Figure 4.14 is a qualitative sketch of the tension forces acting on the wire bead ring. The vector sum of the tension forces around the bead ring must be equal and opposite to the load or normal force that the wheel exerts on the

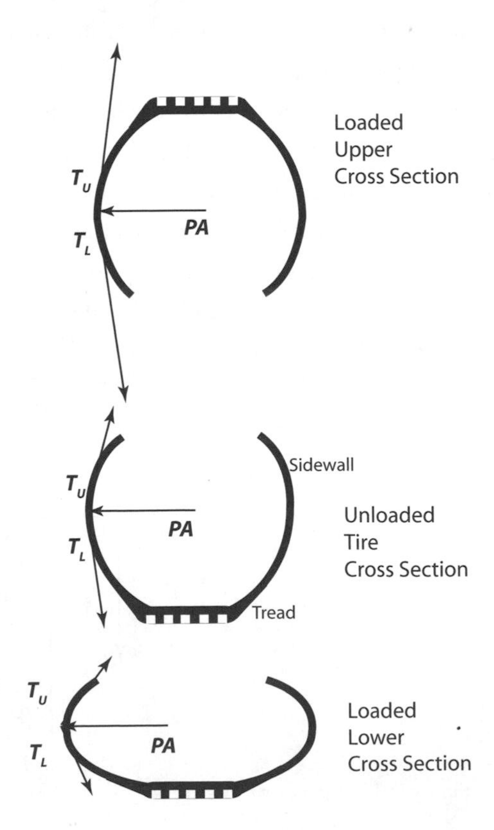

Figure 4.13 Tire cross sections. The middle diagram shows the cross section of the tire below the wheel in an unloaded state. The bottom diagram shows the same cross section when the tire is installed on a car and is at rest on the road surface. We have exaggerated the deformation to emphasize the effect. The top diagram shows the tire cross section for the tire above the wheel when the tire is under load.

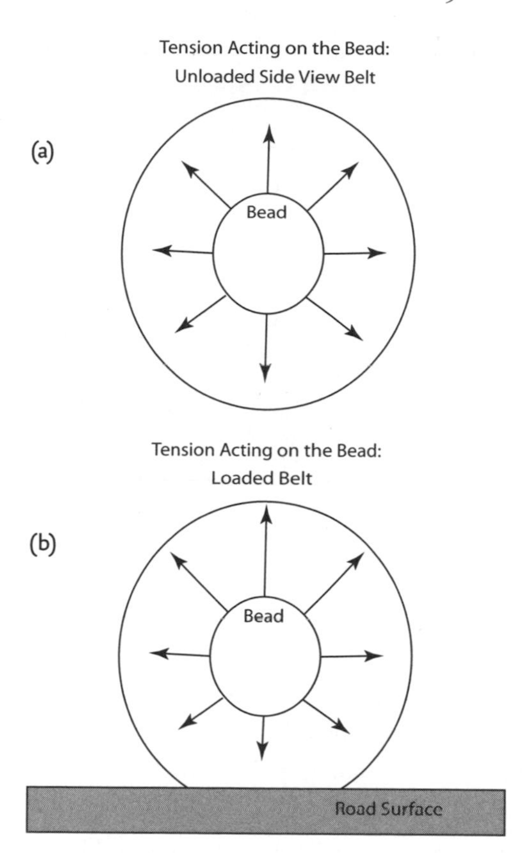

Figure 4.14 Tension forces on the tire bead at rest.

bead ring. (We have neglected the weight of the bead ring because it is tiny compared to the normal force exerted by the wheel on the tire.)

4.9 TIRES UNDER DYNAMIC LOAD

The sidewall tension forces are the communication link between the metal wheels and the contact patch with the road. Consider figure 4.15, which represents a passenger-side drive wheel and tire, where the gas pedal has been mashed to floor. The wheel begins to rotate clockwise while the contact patch is initially stuck to the ground. The sidewall twists like a torsion spring. The tension vectors in the lower tire slew around, pointing to the contact patch, which has fallen behind in the rotation. The sum of the tension vectors sud-

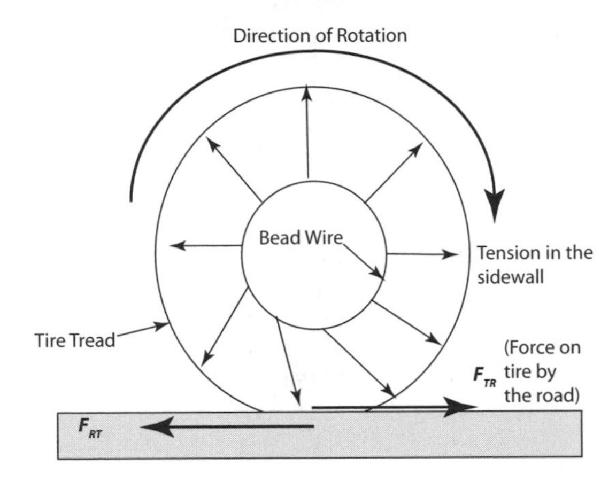

Figure 4.15 Tension forces on a drive-wheel tire bead under acceleration (speeding up).

denly has a net component pointing forward, pulling the wheel and car with it. At the leading edge of the tread before the contact patch, the tire begins to pile up, waiting for the tension to pull the rubber at the contact patch back into place. At the trailing edge, the tread face begins to stretch, reducing the normal force in the trailing edge of the contact patch. This continues until the trailing edge of the contact patch begins to slide. This is referred to as longitudinal slip. The greater the torque applied by the wheel on the tire, the greater the percentage of the contact patch that slips along the ground. This distortion along the contact patch actually allows the tire to rotate at an angular velocity, which is greater than that of a free rolling tire without slip, even though some fraction of the contact patch is not slipping. The deformable tire surface allows simultaneous static and kinetic friction. Describing the coefficient of friction is going to be complicated.

Our driver now shifts to the brakes and the wheel rotation rate begins to slow as shown in figure 4.16. The momentum of the car tries to keep the tire rotating at a rate commensurate with the original speed. The contact patch rotates faster than the wheel, pulling the sidewall backward and rotating the tension vectors in the lower tire toward the back of the car. The leading edge of

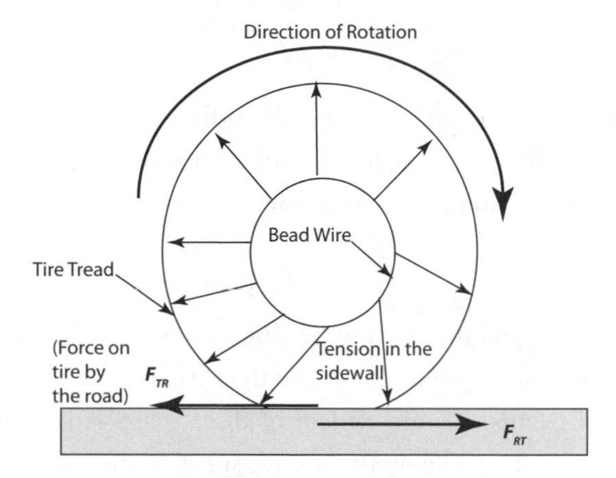

Figure 4.16 Tension forces on a drive-wheel tire bead under braking.

the tread in front of the contact patch is stretched, reducing the normal force at the front of the patch, initiating slip near the front. At the trailing edge of the contact patch, the tread begins to bunch up and compresses. Finally, we see that the sum of the tension vectors produces a net backward force on the bead and the wheel, slowing the car.

We can actually observe the lines of tension that develop in the sidewalls of drag race tires. Drag race tires are unique in that the inflation pressures are low and the sidewalls are soft. This allows, at high angular velocities, the sidewalls and tread to deform and increase the effective radius of the tire. High-speed photographs of the tires at launch show the development of wrinkles that parallel the lines of tension in the lower sidewall.

4.10 CONTACT PATCH FRICTION

4.10.1 Friction in Introductory Physics

In introductory physics, we teach a basic model of friction that has a set of characteristics based on observations. In classical friction, we have either static friction, where the object is at rest relative to the second surface, or kinetic friction, where the contact surfaces move with respect to each other. Kinetic friction is equal to the product of a kinetic coefficient of friction, μ_k, times the normal force, F_N:

$$f_k = \mu_k F_N . \tag{4.7}$$

The kinetic coefficient of friction depends on the materials that make up the two surfaces. For the static case, friction is less than or equal to the static coefficient of friction, μ_s, times the normal force:

$$f_S \leqslant \mu_S F_N . \tag{4.8}$$

It is "less than or equal to" because the friction force will assume any value required to avoid relative motion between the surfaces as long as it is less than $\mu_s F_N$. Classical friction is independent of the surface area of contact. The kinetic case does not depend on the relative speed of the two surfaces.

4.10.2 Friction and Tires

Tires violate all of these characteristics to some degree. This is mostly attributable to the fact that the classical model assumes that the friction is between two rigid surfaces and tires are certainly not rigid. Most introductory texts avoid a discussion of the source of friction. If they do, they make vague arguments about the electrostatic interactions of microscopic asperities (think tiny fingers) on the surface. Friction between the tire and the road shares some of the same origins but is a little more interesting. There are two basic mechanisms involved. The first is adhesion. Visualize a roll of double-sided tape with the outer layer of nonstick separator peeled off. The tire literally sticks to the road. This is not so obvious with street tires, but hot racing tires will pick up all kinds of stuff. You can see this in autocross competition where races are held in large parking lots. Here the cars return to the paddock with tires that look like a Nutty Buddy ice cream cone, with gravel, cigarette butts, and bottle caps stuck to the surface of the tire. As you might guess, adhesion tends to leave a little bit of the tire behind as it rolls across the surface. Go to any race on asphalt or concrete and look just off the racing line. The ground is littered with little balls of tire material. The drivers call them marbles. If you leave the racing line, the marbles adhere to your hot sticky tire and coat it with the cold hard rubber of the marbles. Your car will quickly go sliding off the track because of the loss of traction.

The second mechanism is mechanical keying. In this case, the surface roughness of the road pokes into the surface of the tire as shown in figure 4.17. This roughness is equivalent to a macroscopic asperity, clawing its way along the surface of the tire. The tire surface is elastic in nature and springs back to its original shape. In the extreme case, the mechanical keying tears at the surface of tire. The polymer molecules can slide past each other in the surface of the tire and ultimately chemical bonds are broken. The value of the coefficient of friction will depend on the particular combination of adhesion, mechanical keying, and shredding of the tire surface.

Having an understanding of these basic mechanisms can begin to give us some insight into the way a tire affects handling. For example, what happens in the presence of small amounts of water on the road surface? Adhesion is

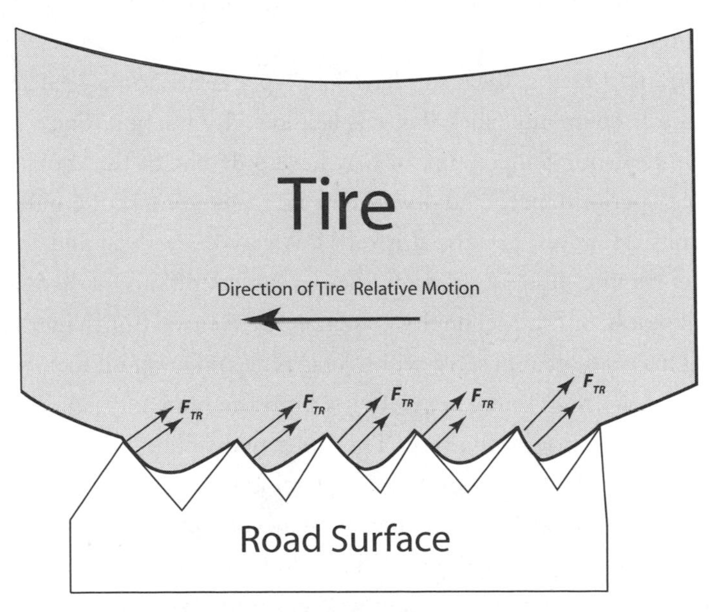

Figure 4.17 Mechanical keying. Idealized road surface roughness interacts with the tire surface.

an electrostatic effect. It depends on the weak induced electrical polarization of the two surfaces. It works well when the two surfaces are in close contact. The presence of liquid separates the two surfaces and breaks the electrostatically induced bonding between the tire and the road. Adhesive friction is significantly weakened. Mechanical keying, on the other hand, is more likely to be effective in damp conditions. The macroscopic roughness of the road surface still digs into the surface of the tire. The normal forces between the sides of the macroscopic asperities and rubber are not significantly diminished by the dampness. Road surface roughness can wear away with time. By moving slightly off the driving line, we can improve traction in damp conditions. When the water becomes sufficiently deep, the tire rides up on a cushion of water, eliminating mechanical keying. The tire is now hydroplaning and traction drops dramatically. The tread on the surface of the tire gives the water an escape route as it is compressed between the tire and the road. The tire and tread act like a pump, removing the water from the contact patch and restoring traction.

In Formula 1 racing, the teams have four types of tire available at any race. First is a soft compound slick that has fabulous dry traction. Once this tire comes up to racing temperature, it may have only one or two "golden laps" of ultimate performance. The driver must be ready to make the most of the opportunity. However, because it is soft, it is easy to overheat and shred the surface of the tire. It has a short life span in hot conditions, perhaps as little as 50 miles or less. The second dry condition tire is made from a harder compound. This is a more robust tire with a longer life and lower ultimate traction. Formula 1 rules require the teams to use both the hard and soft compound tire during the course of the race. The "intermediate wet" is the third type of tire. Water on the track cools the tires, forcing the designers to move to an even softer compound. The intermediates have an optimal temperature performance below that of the dry tires. The addition of a shallow tread pattern improves its ability to combat standing water on the track. The drivers move slightly off the normal driving line to avoid oil floating up off of the track surface. Paul Haney, in his excellent book *The Racing and High Performance Tire*, points out that what the drivers believe to be the effect of oil on the racing line is in reality a mechanical keying improvement. On the negative side, tread reduces the contact area and gives the resulting blocks of rubber the opportunity to squirm. As long as the track remains wet, all is well. As soon as the track begins to dry, the tire temperatures rise as a result of squirm. The tires quickly overheat and traction fades. The drivers begin to weave about the track looking for puddles of water to cool their tires and restore traction. The final type of tire is the "full wet" or "monsoon tire." Deep tread and ultra-soft, low-temperature compounds characterize these tires. The tread patterns pump hundreds of gallons per minute out from between the tire and the road. Tire manufacturers at this level use cutting-edge computer techniques, such as computational fluid dynamics and finite element analysis, to design the tread patterns. As you might guess, the lifetimes of these tires are very short when the track begins to dry.

4.11 MATERIAL PROPERTIES AND TESTING

Tire compounds are one of the black arts of modern technology. Manufacturers mix polymers derived from natural sources with artificially produced polymer compounds, carbon black, silica, antioxidants, oils, fillers, sulfur for polymer cross-linking, and release compounds. The release compounds free the tire from the mold that will give the tire its final shape. Ultimately the mold is used to cook the tire in a process called vulcanization. Manufacturers don't like to share their exact compounds or their exact process. We do know that they will test the mechanical strength, the elastic and inelastic deformation, and the changes in these properties with temperature.

The most common mechanical test for solid materials is a measurement of the Young modulus of elasticity. In this test, a long thin sample is placed under tension and the resulting deformation is measured. Figure 4.18 shows the geometry of a typical test and the resulting stress-strain graph.

We define stress as the applied force, F, divided by the initial cross-sectional area, A_0. It has units of pounds per square inch or newtons per square meter (the pascal), the same units as pressure. Strain is the change in length, ΔL, divided by the initial length, L_0. This ratio is dimensionless. The stress versus strain curve is divided into two regions, the elastic region and the plastic region. In the elastic region, the response is springlike. If the stress is relieved in the elastic region, the material returns to its original length. The slope

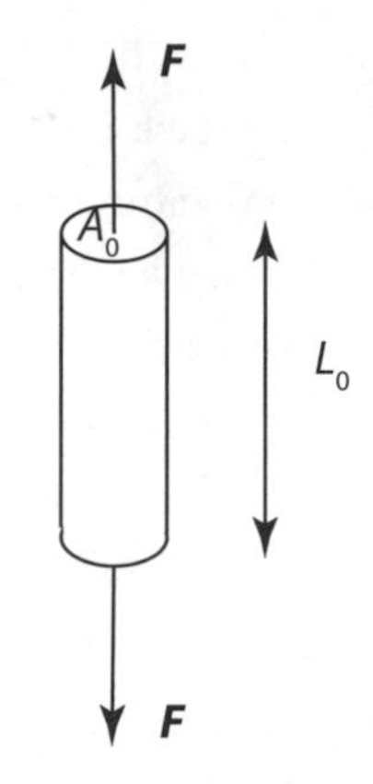

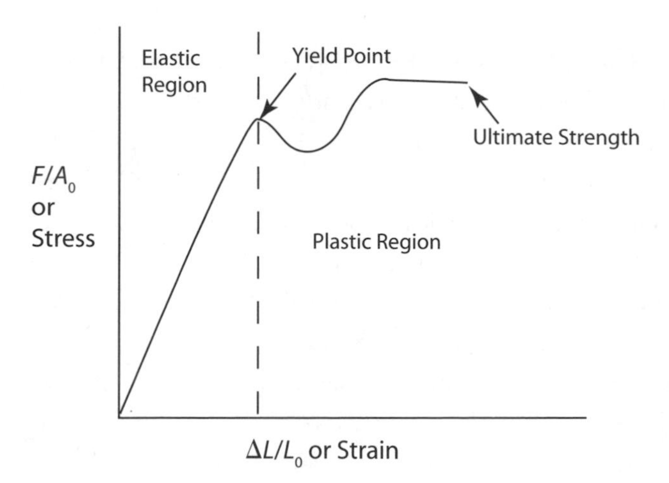

Figure 4.18 Material under tension.

of the curve in the elastic region is called the Young modulus and has units that are equivalent to pressure. This test is used to evaluate metals where the sample is typically crystalline or polycrystalline. A perfect crystal is one where a pattern of atoms is perfectly repeated in a constant pattern filling the entire sample. In a polycrystalline material, the sample had many regions of repeated atomic order. The regions meet at disordered surfaces, called grain boundaries. The yield point in the test is the transition between the elastic and plastic regions. At the yield point, the sample begins to deform permanently. In crystalline or polycrystalline materials, the crystals begin to slip at defects and along grain boundaries. This deformation can reduce the applied stress. In rubber compounds the polymer chains begin to slide past each other at the yield point. Cross-linking between the polymer chains can begin to break. If we continue to increase the applied force, the sample begins to significantly reduce its cross-sectional area. In metals, the defects move through the material until they are pinned in place and the sample begins to stress-harden. The stress reaches a maximum value, called the ultimate strength. Further attempts to increase the stress will fracture the sample. Depending on how brittle or ductile the material is, the exact shape of the curve can vary.

In general, polymers are not crystalline; they are amorphous. In polymers, the repeated structure is limited to the large molecules that make up the chain. Typically, we find the long molecules packed together in a random jumble. The molecules may be cross-linked by bonds, which make the polymer stronger, or they can simply be tangled together. A characteristic temperature, called the glass transition temperature, T_G, is one of the parameters used to describe the behavior of the polymer. Above T_G, the polymer is quite ductile and easily deformed. Below T_G, the polymer hardens and flexibility is lost. In the ductile phase, the yield point is not well defined. In fact, if the stress is held constant for any extended period, the length will continue to grow, even in the elastic region. Initially the polymer unfolds and straightens out, but with sufficient thermal energy the polymer strands begin to slip past each other. With sufficient energy, bonds can be broken. This kind of behavior is similar to a viscous flow in liquids. When the stress is relieved from the polymer, its length begins to shrink but never returns to its original length. It retains a permanent offset from L_0. This type of behavior is called creep and can manifest itself for tires in many ways.

After each track period, the tires undergo a heat cycle. The tires are deformed to some degree. If you park your car with hot racing tires and allow them to cool under load, you will feel the resulting flat spot the next time you move the car. The addition of heat that comes from driving will eventually restore the circular shape. We should point out that this is different from a flat spot that results from locking the brakes and skidding a tire. A skid on a hot racing tire removes a large amount of material. Viscous flow of the rubber cannot restore such a flat spot. High amplitude vibrations can build up if you continue to drive with this type of flat spot. The 2007 Formula 1 world champion, Kimi Raikkonen, learned this the hard way earlier in his career. Late in the race at the 2005 European Grand Prix at Nürburgring, he flat-spotted a tire entering a corner. Several laps later, on the final lap of the race with Kimi in the lead, the resulting vibrations grew so severe that the carbon fiber suspension on his right front wheel shattered, costing him the race.

The tire compound will be subjected to creep tests, cyclic stress tests, and thermal mechanical analysis to find elastic modulus (stress/strain) and loss (lost energy converted to heat) and to determine the glass transition temperature where the rubber loses its flexible properties. Material testing plays a vital role in tire development.

4.12 LONGITUDINAL FORCE AND LONGITUDINAL SLIP

When you step on the gas pedal, the metal drive wheels begin to turn, which rotates the bead and inner radius of the tire. As was shown in figure 4.15, tension forces in the sidewall pull the contact patch of the tire, and in turn the road, toward the rear of the car. By Newton's third law, the road pushes forward on the contact patch and the car accelerates forward. The harder the driver steps on the gas, the greater this tractive force. At the trailing edge of the contact patch, the tension pulls upward on the tread, reducing the normal force and stretching the face of the tire. The tire will begin to slip across the surface of the road at this edge. On the leading edge of the contact patch, the tension is pulling the tire downward, causing it to collide with the road surface. The normal force rises at the front from this collision, and the face of the tire begins to bulge a little. At the leading edge of the contact patch, the tire does not slip and remains in static contact with the road. Thus, if we follow a piece of tread through the contact patch, it transitions from not slipping to slipping at some point along the length of the patch. The percentage of the tire that is in slip is related to the amount of tractive force. Figure 4.19 shows a typical curve shape for tractive forces as a function of longitudinal slip. The initial linear region is typical of classic static friction where slip is not a factor. As slip becomes significant, more of the contact patch has transitioned into traditional kinetic friction. The tractive force peaks and decreases as more

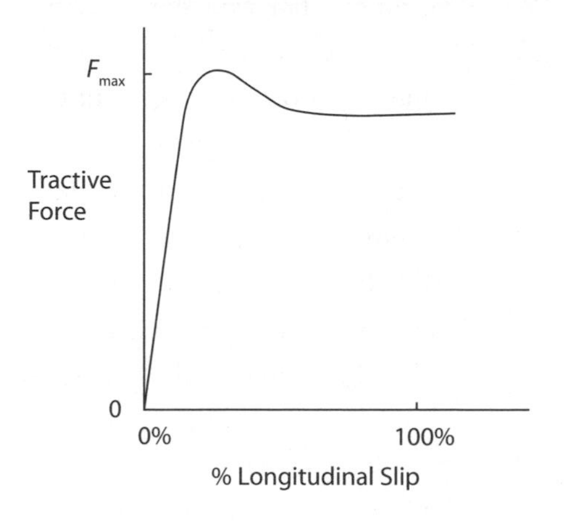

Figure 4.19 Speeding up. Tractive force as a function of longitudinal slip.

and more of the tire transitions into slip. The rate of falloff of the tractive force decreases as a majority of the tire is in slip. A similar curve exists for slip under braking.

It is not surprising that some degree of slip is associated with any degree of acceleration. In essence, the sidewall of the tire acts as a spring. By Hooke's law,

$$F = -k\Delta x ,$$

some degree of deformation of the spring, Δx, is required for the spring to exert a force, F.

It is not obvious at first glance, but any slip at the contact patch while speeding up allows the angular velocity of the wheel to be greater than that of a free rolling wheel moving at the same speed. We will avoid this interpretation for the time being, but there are some circumstances where this interpretation is useful.

Vertical tire load, which is equal to the normal force at the road-tire interface, affects the maximum tire longitudinal force. Equations (4.7) and (4.8) from general physics class describe a linear relationship between the normal force and the maximum friction force. If we plotted the maximum frictional force as a function of the normal force, we would have a straight line with a constant slope equal to the coefficient of friction, μ_s. For most rigid materials,

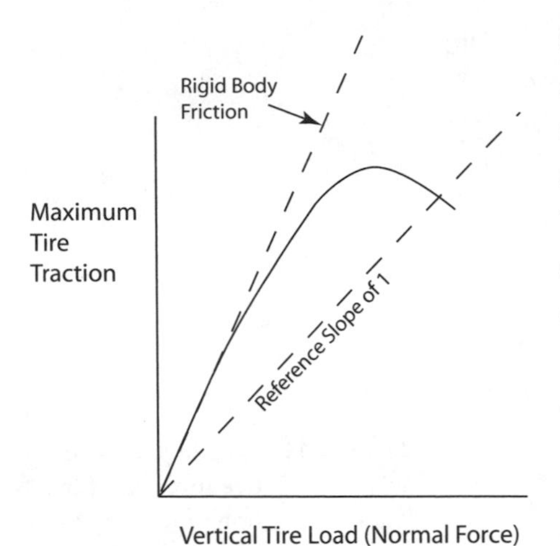

Figure 4.20 Maximum longitudinal force as a function of vertical tire load for a typical racing tire.

the coefficient is less than 1. For example, brass on steel is about 0.5, Teflon on steel is about 0.04, and wood on wood is about 0.4. We know from chapter 1 that for street tires the peak of the static coefficient of friction is around 1.0. For racing tires the coefficient can exceed 1.0. Figure 4.20 is a plot of maximum longitudinal force as a function of vertical tire load for a typical racing tire. The slope of this curve can be thought of as a coefficient of friction as well and is not constant. It tells us, for a given load condition, how much additional traction we will gain for a small increase in the vertical load. There are two straight lines on the plot as well. The upper line is tangent to the initial slope of the curve. The slope decreases with increasing load. The second straight line is a reference line with a slope of 1. Slopes less than this line are typical of friction between rigid objects. Slopes greater than this line are typical of racing tires at normal

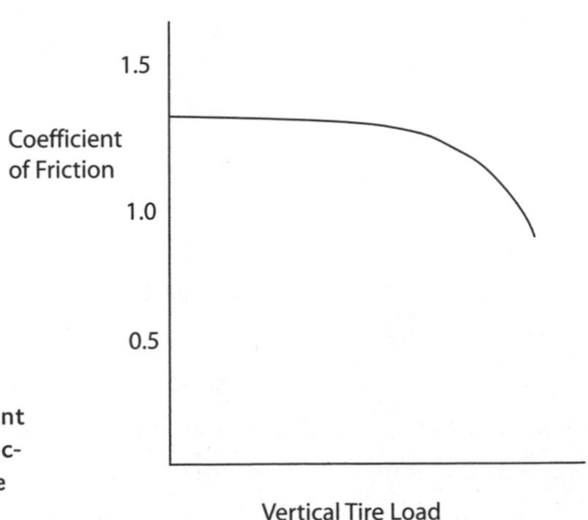

Figure 4.21 Effective coefficient of friction as a function of vertical tire load.

operating temperature and moderate load. We saw in chapter 1 that drag slicks launching on the hot sticky rubber laid down by a previous burnout could produce a coefficient of 4.0 or greater! The important idea to take away is that transferring more load to an overloaded tire does not improve traction. Figure 4.21 is a plot of the effective coefficient of the total load acting on the tire.

4.13 LATERAL FORCE AND THE SLIP ANGLE

To develop our understanding of longitudinal force in the tire, we earlier treated the tire as a spring that connects the wheel to the contact patch. It should not be a surprise that lateral forces have a similar analogy. When we turn the steering wheel, the metal wheel and the tire bead twist in the direction that you have turned. The contact patch is initially constrained to roll in the direction the car is traveling. The tire is twisted like a torsional spring. The torque on a torsional spring is proportional to the angle of displacement, $\tau = -\kappa\alpha$, where τ is the torque, κ is the proportional constant, and α is the angle. The tire is not quite analogous to the torsional spring because the rolling of the tire allows the rubber to slip back into place as it passes through the trailing end of the contact patch. Figure 4.22 shows the tire heading at an angle α with respect to the instantaneous velocity of the car (direction that the car travels). The twist gives rise to a lateral force on the wheel. Figure 4.23 shows a similar drawing following the motion of the centerline of the tire as it passes through the contact patch of the car. The distortion in the

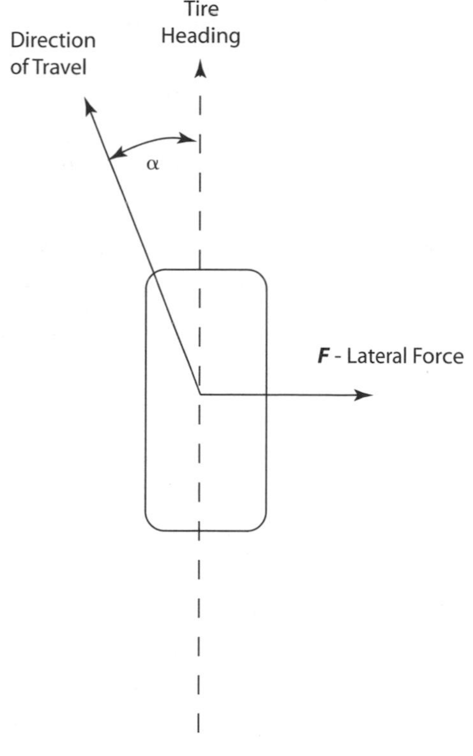

Figure 4.22 Tire slip angle (α) defined.

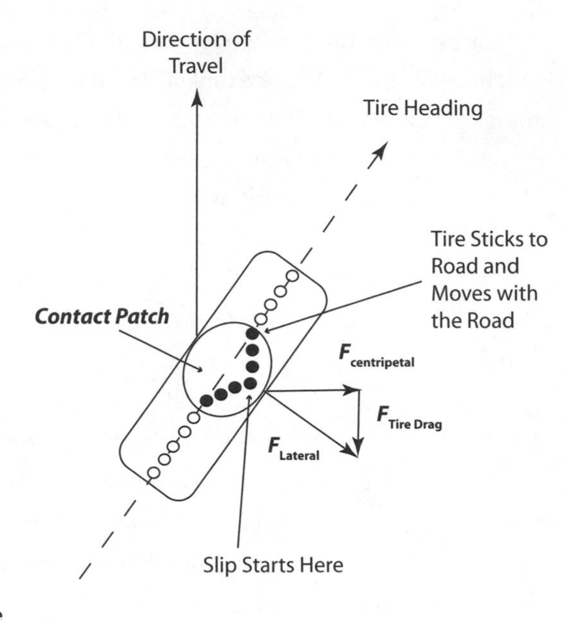

Figure 4.23 Distortion in the contact patch gives rise to the lateral force.

figure is exaggerated to clarify the effect. As the centerline enters the contact patch, it is constrained to move with the road because it is not slipping. This movement distorts the contact patch sideways with respect to the tire. The bigger the distortion, the larger the tension forces in the tread face. As we move toward the rear of the contact patch, the normal force decreases and the tread begins to slip. The total effect is to produce a triangular-shaped distortion in the contact patch that gives rise to the lateral force.

The lateral force, $\boldsymbol{F}_{\text{Lateral}}$, is perpendicular to the wheel. This force, as shown in figure 4.23, can be decomposed into two components. The first is centripetal force, $\boldsymbol{F}_{\text{centripetal}}$, which causes our circular motion. The second is $\boldsymbol{F}_{\text{Tire Drag}}$, which is referred to as the induced tire drag. The act of turning the car creates a force that slows the car. The bigger the slip angle, the greater the induced tire drag. Do not confuse this with the aerodynamic term. A separate aerodynamic drag for tires exists, which is also very significant for an open-wheeled car.

Figure 4.24 shows the behavior of a typical plot of lateral force as a function

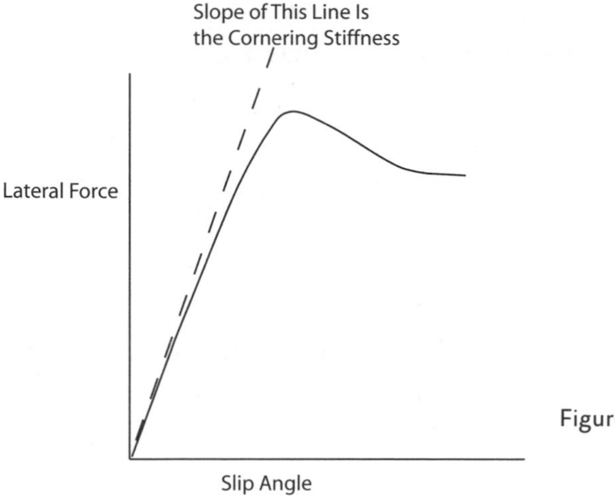

Figure 4.24 Typical lateral force as a function of slip angle.

of slip angle. Again, because this is a spring system, some distortion or slip angle is required to generate a force. The initial slope of this curve is called the cornering stiffness and has units of lbs per degree. The cornering stiffness is a function of tire design, internal pressure, and tire vertical load. Cornering stiffness is an important factor in handling performance on corner entry. Notice that at some critical slip angle the slope of this plot falls off significantly. Ultimately, the slope actually changes sign. Looking at this plot, it is easy to envision a situation where further turning of the steering wheel produces less turning force and more induced tire drag. This is one of the factors leading to understeer.

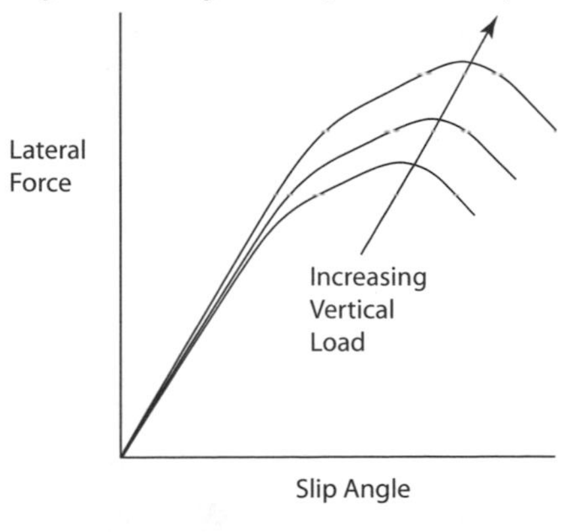

Figure 4.25 Lateral force as a function of slip angle for the same tire under identical conditions.

One of the things that figure 4.24 does not tell us is what happens when load transfer takes place under cornering. Figure 4.25 shows lateral force as a function of slip angle for the same tire under three different vertical loads. The initial cornering stiffness increases slightly and remains linear longer with increasing vertical load. We know that the pressure in the contact patch is controlled predominantly by the inflation pressure in the tire. The increased vertical load increases the contact patch area as the bottom of the tire flattens. This effect spreads the distortion in the contact area over a larger area and, presumably, a smaller percentage of the patch ends up slipping over the road surface.

In summary, regardless of the vertical load, when the slip angle becomes too great, the lateral force falls off and the induced tire drag becomes large. Any further turning of the wheel simply acts to slow the car. However, any driver action that transfers vertical load to the tire will improve the cornering performance.

4.14 ALIGNING TORQUE

How does the driver know when the slip angle is too great? To some degree, he or she can feel the yaw rate (rate of turning left or right) with their inner ear or through forces in their seat. A lot of drivers and instructors talk about having a well-calibrated seat! However, these changes are subtle. A better feedback mechanism is a well-made steering system using something called a self-aligning torque. The idea is to have the lateral force at the contact patch create a torque, which drives the steering wheel back toward the neutral position where the car heads straight.

Figure 4.26a shows the side profile of a tire with a vertical steering axis. The wheel, hub, and steering knuckle all pivot about the kingpin. The axis intersects a point on the ground next to the center of the contact patch. The lateral force generated in the contact patch does not generate a torque about the steering axis in this orientation. In figure 4.26b the steering kingpin axis is tilted with the top backward. This angle of the kingpin with respect to the vertical is called the caster angle, and in this orientation it is positive. If we project the steering axis downward until it intersects the ground, we define the "Dave point" (named by noted magazine journalist Dave Coleman). The horizontal distance from the Dave point to the center of the contact patch is called the mechanical trail. The lateral force in the contact patch generates a torque about the steering axis and is a function of the mechanical trail and

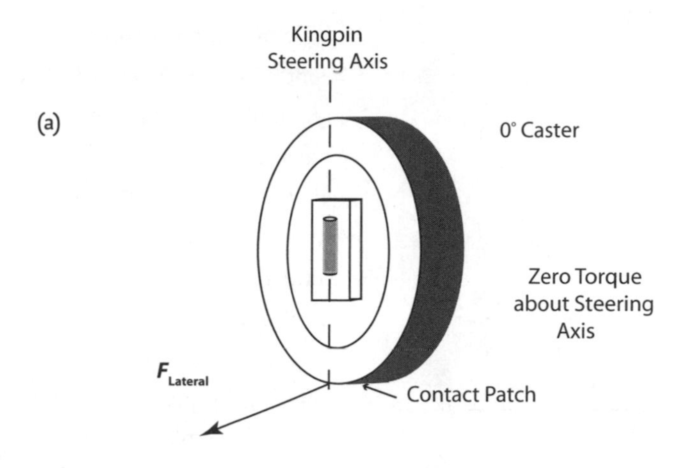

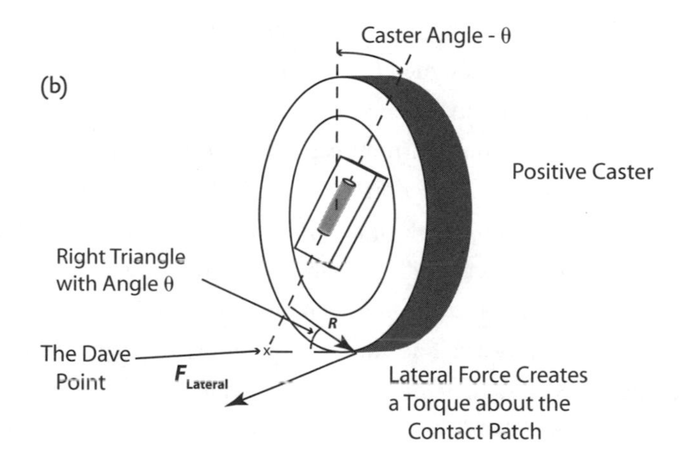

Figure 4.26 Steering kingpin angle, or caster, and aligning torque.

the lateral force. Because of distortion in the tire under load, the lateral force on average does not act exactly at the center of the contact patch. This shift in the point of action of the lateral force is called the pneumatic trail. The total aligning torque from the caster is the sum of the mechanical and pneumatic trails times the lateral force times the cosine of the caster angle. We can apply the small-angle approximation, where $\cos(\theta) \approx 1$. The caster angle is typi-

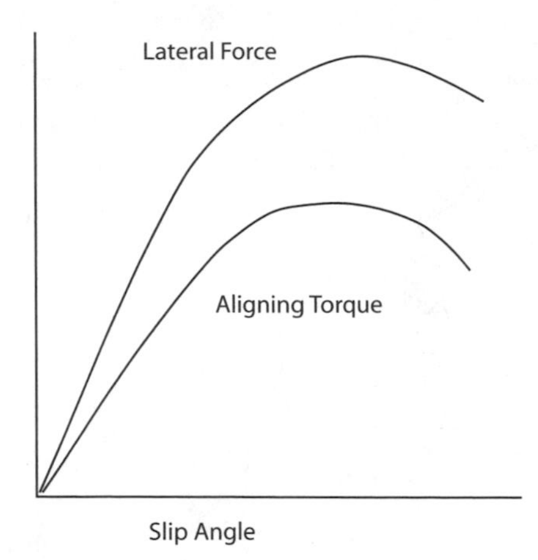

Figure 4.27 Lateral force and aligning torque as a function of slip angle.

cally only a few degrees, making our approximation accurate to within a few tenths of a percent. The resulting aligning torque rotates the tire to align with the instantaneous velocity vector. The wheels and steering return automatically to the neutral position.

The driver must overcome the aligning torque to turn the car. Figure 4.27 is a plot of both lateral force and the aligning torque as a function of the slip angle. Typically the aligning torque falls off a little faster than the lateral force. In a well-designed car, the driver feels the falloff of aligning torque as an indicator of falling lateral force. Any further cranking of the steering wheel gains no additional turning capability. In fact, further turning of the wheel increases induced tire drag, slowing the car. The front tires begin to skid. We are deep into understeer, and the car simply refuses to turn. The driver must take some other action to make the corner. Turning the steering wheel further will cut the lateral force even more. Lifting off the throttle will transfer vertical load forward, and slightly unwinding the wheel will restore rotation of the front tires, allowing them to stop skidding. The optimal racing line is lost, but at least the driver regains control of the car. Turning past the peak in the lateral force curve is a very common mistake in autocross racing, and unwinding the steering is very counterintuitive. They say knowledge is power. In this case, knowledge is speed.

4.15 SUMMARY

In this chapter we developed a method to find the center of gravity for the car. We applied Newton's laws and calculated the amount of vertical load transfer that occurs when the car accelerates in a horizontal plane. We reviewed basic tire and wheel construction and identified the important dimensions for each.

We learned that as a deformable body, the rules for friction differ from those we applied to rigid bodies in general physics class. We studied the parameters that affect tire performance. Our findings are summarized as follows:

- The longitudinal and lateral center of gravity can be determined using four scales beneath the tires on a flat horizontal surface.
- Finding the height of the center of gravity requires that we elevate one end of the car and compare the results with the level horizontal case.
- A longer wheelbase reduces vertical load transfer when the driver steps on the gas or the brake.
- A wider wheelbase lowers vertical load transfer when turning.
- A lower center of gravity lowers vertical load transfer under all accelerations.
- The ratio of the wheelbase over twice the height of the center of gravity is called the static stability factor (SSF). The SSF times the acceleration of gravity should be much greater than the lateral acceleration ability of the tires.
- If we multiply the tire pressure times the contact patch area for the four tires, the product is equal to the weight of the vehicle in equilibrium.
- Increasing the vertical load on a tire increases the contact patch area but does not significantly change the tire pressure.
- Tire pressure pretensions the tire sidewall. Vertical load reduces sidewall tension in the lower part of the tire and raises the tension in the upper part of the tire.
- Tension in the tire sidewall is the mechanism by which the contact patch forces are transmitted to the tire bead, the wheel, and the car.
- Tire contact patch friction is due to both adhesive forces and forces that arise as a result of mechanical keying.
- A tire contact patch can simultaneously stick and slip because the surface deforms. Longitudinal slip initially accompanies a rise in tractive forces, followed by a drop in tractive force with additional slip.
- Maximum tire traction is less than linear with vertical tire load. Additional vertical load can reduce traction. This is equivalent to defining a coefficient of friction that decreases with vertical load.

- When turning, the front tires must be turned to an angle beyond the direction of the instantaneous velocity. This is the slip angle.
- A tire slip angle generates lateral force and an induced tire drag.
- Lateral force is less than linear with slip angle. Additional slip angle can reduce the lateral load.
- Caster angle creates an aligning torque that restores the tires and the steering wheel to the neutral position.

Chapter 5

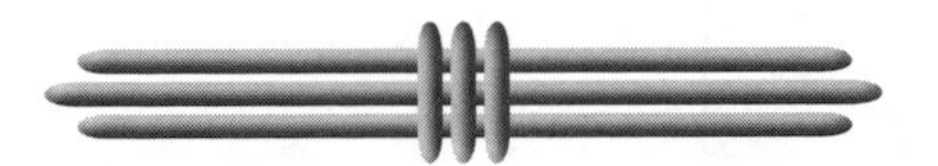

Steering and Suspension

We now have some criteria for choosing a racing line, and we understand a little more about how tires work. It is time to tune the handling of our car. To understand handling corrections, we need to integrate what we have already learned with new ideas from suspension and steering design. We will start with steering basics, steering problems, and common suspension layouts. We will review the various types of springs and dampers, as well as the basic physics of damped spring systems. We will then be in a position to understand how load transfer affects handling. Finally, we will put this all together to correct handling problems.

5.1 MORE ON STEERING

Before we can understand the origins of oversteer and understeer, we need to know more about steering geometry. In almost all passenger vehicles the rear tires do not turn. Their orientation is fixed with respect to the longitudinal axis of the car. At low speeds, the slip angles of the tires are small and we can ignore them. In a constant-radius turn both of the rear tires will trace out a circle on the ground. The outer wheel must rotate faster in the turn, and the radius of the turn for the outer wheel will be larger than the inner wheel by an amount equal to the wheel track of the car, t. The center of curvature (the

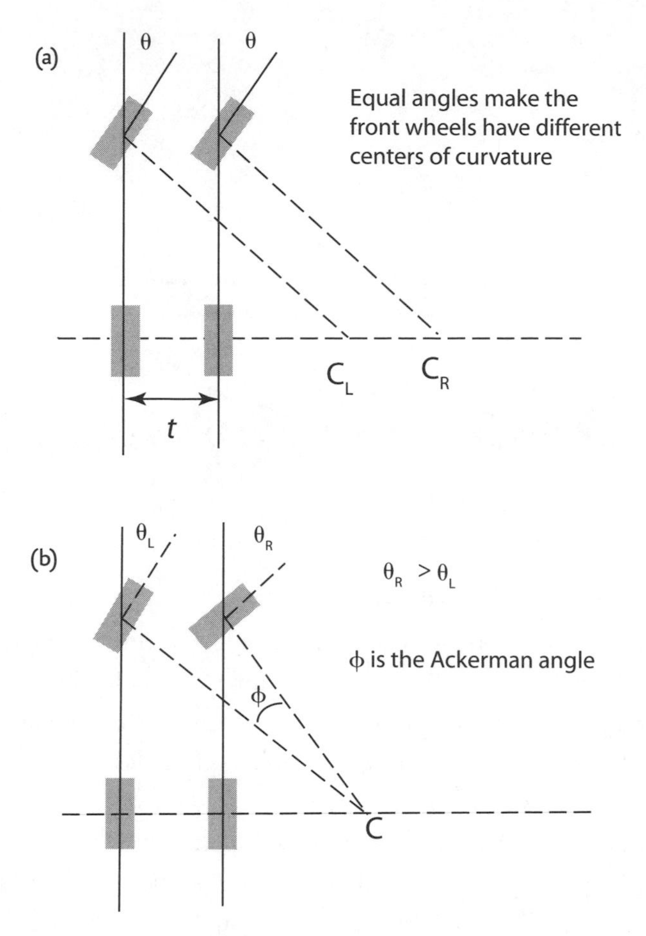

Figure 5.1 Ackermann steering: low speed.

center of the circle) must lie on the axis of all of the wheels if the tires are to roll smoothly. Figure 5.1a shows the situation if both the left front and right front wheels turn the same amount. The dashed lines are projections of the axes of the wheels. In this situation the tires will continuously fight each other because each is trying to follow a different path.

If we adjust the angle of one the front tires, we can force the three dashed lines to intersect at a single spot. This creates a single center of curvature for all wheels. Figure 5.1b shows this arrangement. We call the difference between the angles of the two front wheels Ackermann steering. While the concept of

Ackermann steering and Ackermann geometry is famous in automotive engineering, I had not heard the story of Rudolph Ackermann until I read *Automotive Engineering Fundamentals* by Richard Stone and Jeffery Ball. It seems that Ackermann was not an engineer, but an English patent lawyer, hired by a German engineer named Lankensperger in 1817. Lankensperger had solved the steering problem for wagons and carriages, and Ackermann captured the glory. Some things never change. Prior to this time, the front wheels were attached to a single axle. To steer, the entire axle and wheel assembly turned as shown in figure 5.2a. This worked at low speed but failed at high speed and

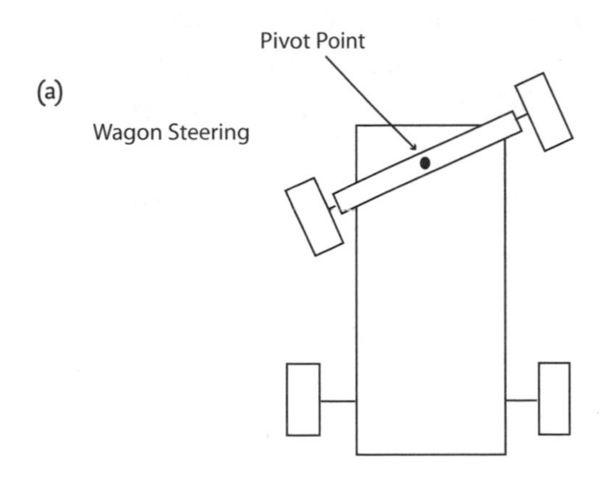

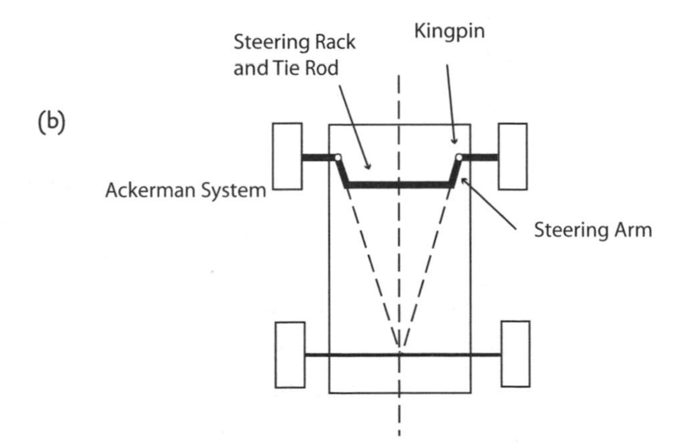

Figure 5.2 Lankensperger steering (as patented by Ackermann).

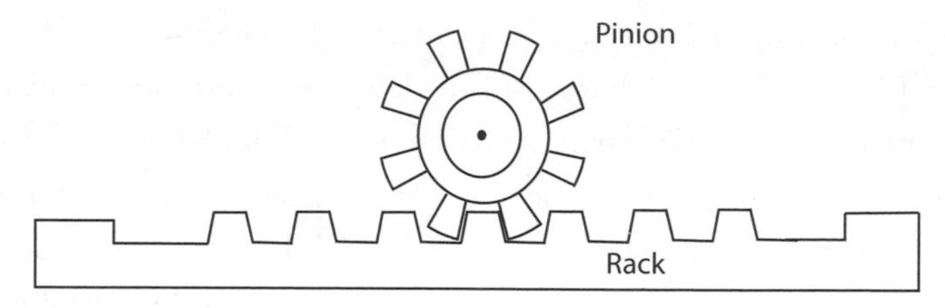

Figure 5.3 Rack and pinion steering.

under heavy loads. By the time cars arrived on the scene, they completely overpowered the front fixed axle system. The Lankensperger solution is shown in 5.2b. The two front wheels turn independently, and each is steered by a short control arm that trails aft from the wheel and cants toward the centerline of the car. If we draw a line along each of the two control arms, the lines intersect at the midpoint of the rear axle. A single horizontal member connects the two control arms. The resulting tire angles produce the desired smaller turn for the outside wheel. The Ackermann geometry is not perfect for all turns, but tire slip angle can accommodate minor corrections.

The horizontal member that connects the two steering control arms is typically three pieces, a tie rod on both ends and a steering rack in the middle. The tie rods have a ball joint at each end that provide a pivoting joint. The steering rack is typically a rack and pinion as shown in figure 5.3. The pinion is attached to the steering column and converts the rotation of the steering wheel into linear motion of the rack. The diameter of the steering wheel, the diameter of the pinion, and the length of the control arms determine the leverage that the driver has to apply to turn the tires.

5.2 BICYCLE MODEL: OVERSTEER AND UNDERSTEER

As speed increases, the radii of turns employed tend to grow larger. Eventually the wheel track, the distance between the inside and outside wheels, becomes less significant. We can ignore the track and treat the car as a single set of front and rear wheels, referred to as the bicycle model. The bicycle model allows us almost immediately to see the effect of slip at the two ends of the car.

Consider a car traveling in a constant-radius circle. For this case, the front

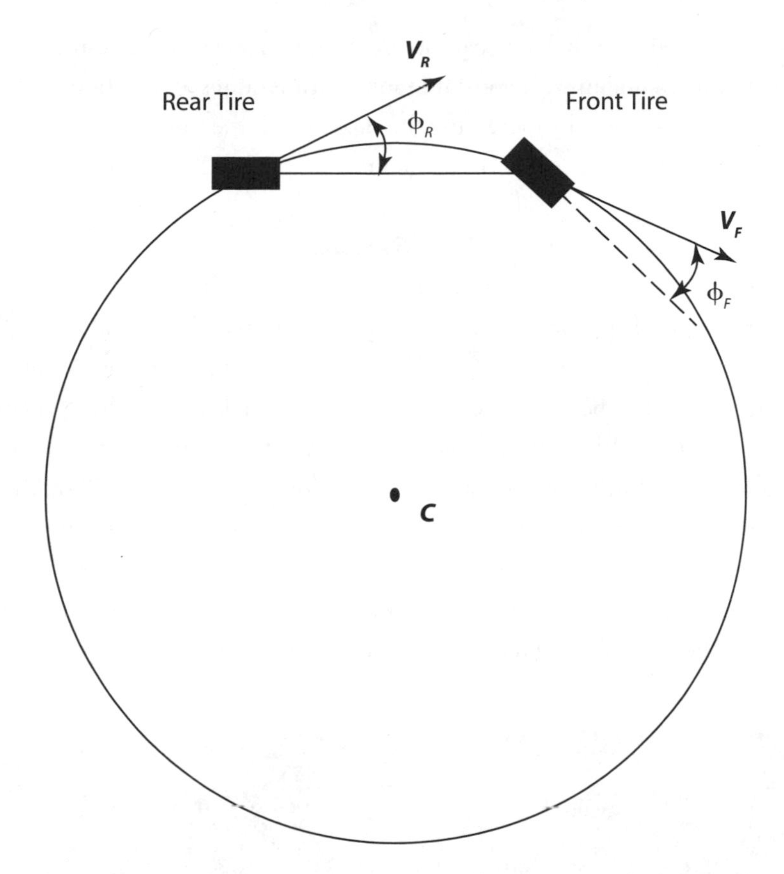

Figure 5.4 Front and rear slip angles in the bicycle model.

and rear tires must both have a velocity vector tangent to the circle as shown in figure 5.4. The front tire is turned slightly inward from the front velocity vector, $\mathbf{V}_F$, and creates a front slip angle, ϕ_F. The rear is constrained to point along the axis of the vehicle. The angle between the rear tire velocity vector, $\mathbf{V}_R$, and the longitudinal axis of the vehicle is the rear slip angle, ϕ_R.

5.2.1 Neutral Handling

If $\phi_F = \phi_R$, the car has neutral or balanced handling. This is the optimal condition. The car is best able to maneuver and has the most overall grip when the car is balanced. This balance can be upset with changes in speed. Increasing speed while maintaining a constant-radius turn requires an additional lateral force. This is attained by increasing the slip angle. If the front and rear tires do not increase the slip angle at the same rate, the car will become unbalanced.

5.2.2 Oversteer

In the oversteer condition, the front slip angle is less than the rear, $\phi_F < \phi_R$. The rear end of the car begins to slide out to a greater turn radius. In the extreme, it leads to a spin. This is a challenging handling condition. NASCAR drivers refer to it as being "loose." On the street, it can lead to entering oncoming traffic lanes. When the car gets loose or oversteers, our natural instinct is to slow the car. Lifting off the gas pedal or stepping on the brake transfers load forward in the car and reduces rear grip. Our natural instincts lead to a spin. Lawyers for automotive companies hate oversteer. The photo below demonstrates oversteer in a 3.0 CSL BMW. You can tell that the car is turning left because the car is leaning hard to the right. But instead of the front tires being

Catching a massive oversteer! Photograph by Clyde Caplan and Alex Teitelbaum.

turned to the left, they are full lock to the right trying to stop the rear end from coming around and passing the front.

5.2.3 Understeer

In understeer the front slip angle is greater than the rear, $\phi_F > \phi_R$. The front of the car begins to drift out to a greater turn radius. In the extreme, the front slip angle is so great that lateral force begins to decrease with increased steering angle. The front end of the car simply refuses to respond to the driver's input. In NASCAR, they call this condition "push" because the front end pushes outward. They also call it being "tight," because the front end feels like it's bound up and unable to respond. On the street, the front end begins to drift into oncoming traffic lanes. Once again, our instincts kick in and we lift off the throttle or step on the brake. In this condition, we get lucky. Vertical load transfers forward, increasing the contact patch area and the overall front grip. The car begins to return to the proper lane. Lawyers for car companies love understeer.

5.2.4 Conclusions

You might think at this point that the lawyers would go for neutral handling since it provides the best overall response. The problem is that conditions can change. Tire pressures can drift or alignment can change with the bump of an unnoticed curb or pothole. Neutral handling has an equal probability of becoming oversteer or understeer. However, if they start with understeer in the design, when changes occur the handling can become worse in understeer or more neutral. The dreaded oversteer is less likely. As a result, almost every car that leaves the factory leaves with built-in understeer. For a track-day car or an autocross car, this is the first thing most drivers try to fix.

This is a good opportunity to consider the centripetal forces involved in making a constant-radius turn. Figure 5.5 takes the previous figure and adds the lateral forces generated by each tire. The lateral force, $\boldsymbol{F}_{\text{lat}}$, is perpendicular to the tire. The centripetal force must be perpendicular to the velocity vector. The lateral force must be broken down into components that are perpendicular and antiparallel to the velocity vector. The perpendicular component is

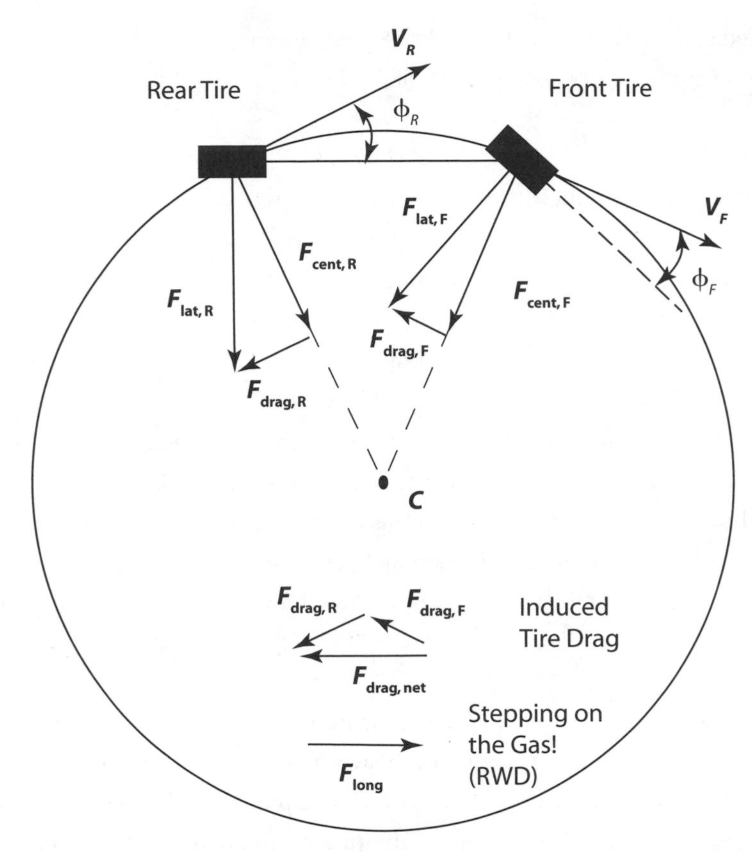

Figure 5.5 Centripetal and induced tire drag forces in the bicycle model.

the centripetal force, $\boldsymbol{F}_{\text{cent}}$, and the antiparallel component is the induced tire drag, $\boldsymbol{F}_{\text{drag}}$. The vehicle is going to slow down as a result of the turn. To maintain speed, the driver must step on the gas and generate a longitudinal force at the rear drive wheels.

It is important to point out that the figures exaggerate the geometry to clarify the situation for learning. The radius of the turn is too small and the slip angles are too big. However, the principles demonstrated are clear and correct.

5.3 WHEEL ALIGNMENT

While we are discussing steering, it is worth taking a moment to consider wheel alignment. Typically three parameters are considered part of the alignment: caster, camber, and toe. We have already touched on caster (fig. 4.26). Caster is inclination of the steering kingpin with respect to the vertical. It is responsible for the self-aligning torque that returns the steering wheel back to neutral following any turning maneuver. When the steering wheel is turned, caster causes the outside front wheel to lean into the turn with negative camber. In general, this improves corner turn-in response.

Toe refers to the angle of the tires with respect to the longitudinal axis of the car. Toe-in is the condition where the front edges of the tires are both turned inward toward the centerline of the car as in figure 5.6. Toe is usually specified as the difference in distance between the front edges of the tires and the rear edges ($D_R - D_F$), and a toe angle of $\theta = 0°$ would appear to be the ideal case in which, in straight-line motion, the tires do not produce any lateral forces.

This is complicated by the fact that most cars have to some degree a non-

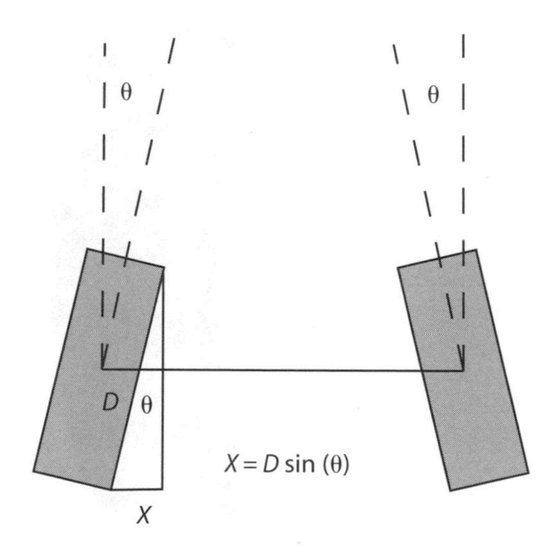

Figure 5.6 Front tire toe-in.

zero scrub radius. The scrub radius is shown in figure 5.7. It is the distance, at the road surface, between the center of the contact patch and the point where a projection of the steering pivot axis meets the ground (the Dave point). Longitudinal forces in the contact patch produce a torque about the steering axis as a result of the lever arm provided by the scrub radius. For example, rolling resistance creates a torque, which tends to push the tire in the toe-out direction. In a rear-wheel-drive car, rolling resistance is the primary concern for longitudinal force. To cancel the effect of the rolling resistance torque, the front wheels are aligned with a slight toe-in. In front-wheel-drive cars, the driving force, sometimes called the tractive force, creates a force that drives the wheels in the toe-in direction. It is common to run a slight toe-out to

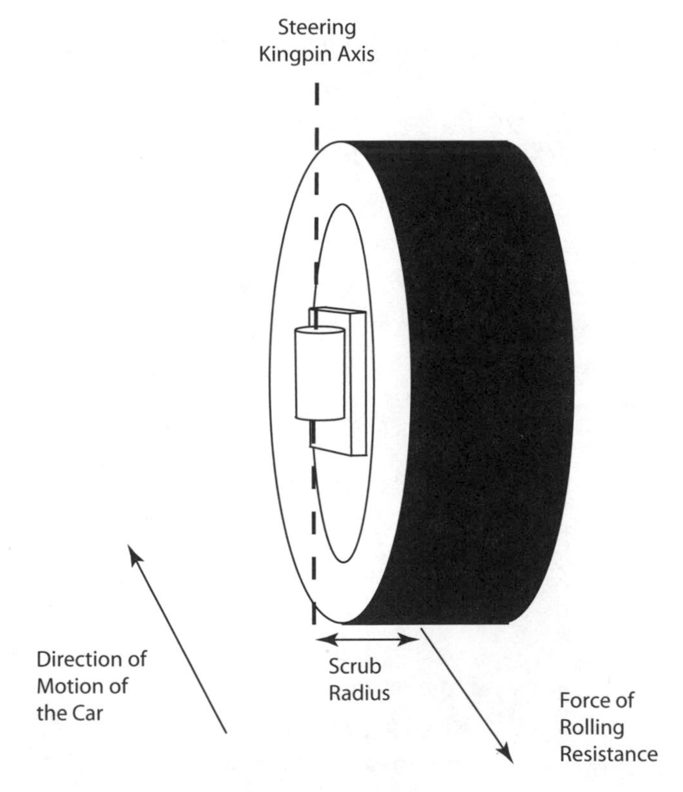

Figure 5.7 Front tire scrub radius.

compensate. The actual toe angle, θ, is usually a fraction of a degree. From a practical standpoint it is difficult to measure a fraction of a degree for a tire angle, which is why they resort to a measurement of the distance between the tires. From figure 5.6 we can see that the toe is equal to $2x$ and $x = D \sin(\theta)$, where D is the tire diameter. Plugging in a typical 24 inch tire diameter and $\theta = 0.25°$, we get 0.23 inches, or just under a quarter of an inch. The Subaru STi toe specification is 0 ± 3 mm, which is roughly one-eighth of an inch.

Improper toe setting is the most likely alignment problem. A tap of the curb while parking or a pothole can knock out the toe setting. Improper toe is also the most likely source of premature wear in your tires. If your tires wear on one side more than the other, your toe alignment is most likely incorrect.

The use of toe-out sets up an unstable equilibrium. When used in the front end in small amounts, it can improve corner turn-in at low speeds. This is a common trick used in autocross, but it's a bad idea for a car that is your daily driver. Imagine dealing with front-end steering instability in the rain as your tires dart from puddle to puddle. Small amounts of toe-out can be used in the rear tires. As you enter a turn, vertical load transfers to the outside tires. The rear is already in a toe-out, and with vertical load, it goes to a large slip angle. This oversteer condition improves corner turn-in, if the driver can control it. Toe-out in general is not a good idea for inexperienced racers.

As a final word about toe, we should discuss bump steer. Bump steer is a condition where the toe angle changes when the suspension moves up and down. In an ideal situation, motion of the suspension should not affect the steering of the car. This is relatively easy to achieve if the steering angle is small when the suspension compresses. By placing the inner tie rod pivot point on the imaginary line that joins the upper and lower control rod inboard pivot points, and doing the same for the outboard pivot points, steering bump will be minimized. This is shown in figure 5.8. When the steering angles are large, this is significantly more difficult to achieve. Design engineers spend a great deal of time trying to minimize the effects of bump steer. Adding aftermarket suspension parts, such as adjustable control arms, frequently introduces bump steer. Correction of the problem frequently requires moving the steering pivot points. Cutting and welding of suspension components is usually outside the

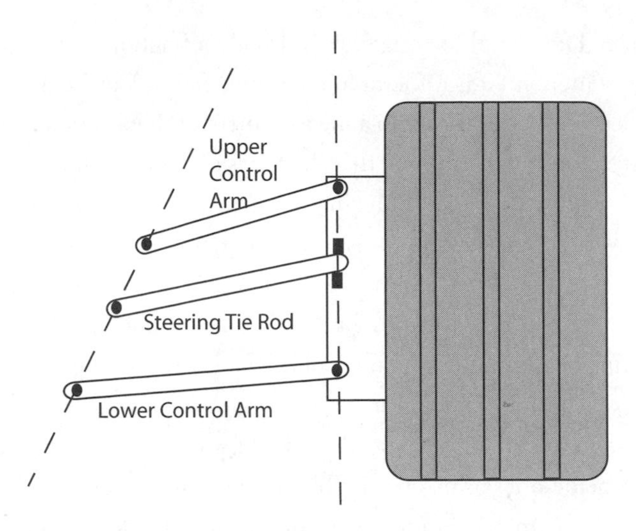

Figure 5.8 A geometry to minimize bump steer.

scope of most weekend warriors. Carroll Smith in his book *Prepare to Win* suggests that shims in the tie rod connecting points can be used to correct the geometry in some cases. If you replace suspension components that change the geometry, take the time to remove the springs and measure the resulting bump steer. Herb Adams in his book *Chassis Engineer* places an upper limit of 0.010 to 0.020 inches of toe change per inch of suspension travel for tolerable bump steer. If you can't achieve this with shims, consider removing the aftermarket part.

The last adjustment parameter we will consider is camber. Camber is the tilt angle of the wheel and tire away from vertical. If the top of the wheel is tilted inward toward the car centerline, the camber is considered negative. A tire lowered onto the ground with an initial negative camber will have a contact patch that is curved, with the concave part of the curve facing inward toward the centerline of the car as shown in figure 5.9. Once the tire begins to roll, the rubber that enters the leading edge of the contact patch will try to move straight backward along with the piece of asphalt that it initially meets. The asphalt will exert an inward-directed friction force, distorting the tire. This inward-directed force is called camber thrust. By symmetry, camber thrust forces from the left and right tires sum to zero until the car enters a turn. On corner entry, the vertical load transfers to the outside tires and the camber thrust becomes unbalanced in favor of the outside tires. This improves corner turn-in. It is common for racers to run 3° or 4° of negative camber.

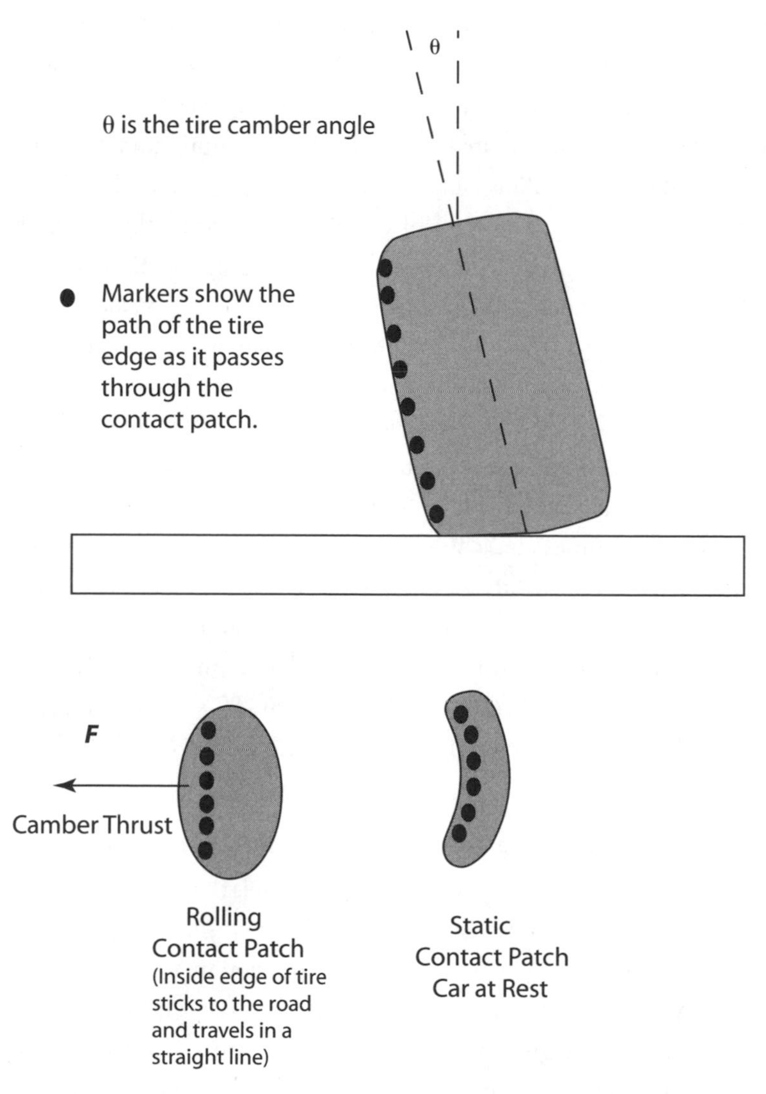

Figure 5.9 Tire camber and the resulting force called camber thrust.

5.4 SUSPENSION BASICS

The suspension has a number of functions and accommodations. To some degree, it should isolate the chassis from bumps and potholes of the road. It should keep the drive and steering wheels in firm contact with the road. It should control the roll, dive, and squat of the body, relative to the wheels. It should dampen out any of the resonant motion of the system. It must accommodate steering of the front tires. It must accommodate the drive axles and brake system, as well as control the reaction to the resulting tractive and braking forces. It must also allow adjustments to the alignment of the wheels.

The metal fixture at the end of the suspension arms that serves as the mounting point of the brakes, wheel, and steering connection is called the knuckle. Milliken and Milliken in their book *Race Car Vehicle Dynamics* point out that the suspension constrains the motion of the knuckle to a single path of motion. Since a free object has six degrees of freedom, three linear and three rotational, the suspension must constrain five degrees of motion. If a connection or link can constrain only one degree of motion, we'll need five links. Many suspension components constrain more than one degree of freedom, so it is sometimes challenging to identify the equivalent to the independent five links. There are literally dozens of suspension types. We will limit our discussion to only a few to learn some of the general principles. We consider two types of independent suspensions, the double A-arm and the MacPherson strut suspensions. We will also consider the NASCAR-style solid rear axle. It's primitive, but an important part of American motorsports.

5.5 DOUBLE A-ARM OR WISHBONE SUSPENSION

The double A-arm or double wishbone suspension is the most common racing suspension. Even NASCAR employs it for the front suspension. Figure 5.10 shows the layout with an upper and lower suspension member in the shape of the letter A. Each A-arm acts as two suspension links, so a fifth link is required to control the final degree of freedom of the knuckle and wheel. Two equal-length A-arms initially seem like a pretty good idea. Motion of the wheel retains the vertical orientation, which maximizes the contact patch. However, when the car body rolls in a turn, the outside heavily loaded wheel is pushed in the direction of positive camber. Positive camber reduces the lateral force, an unintended consequence.

We can correct the camber by employing a shorter upper A-arm. In this situation, vertical motion of the wheel induces negative camber. When the body rolls, the actual wheel camber will remain neutral. Figure 5.10 shows a sketch of the unequal-length double A-arm suspension.

What does the motion of the wheel look like if the tire runs over a bump? Figure 5.11a is a view from the rear of the car showing the left wheel, a simpli-

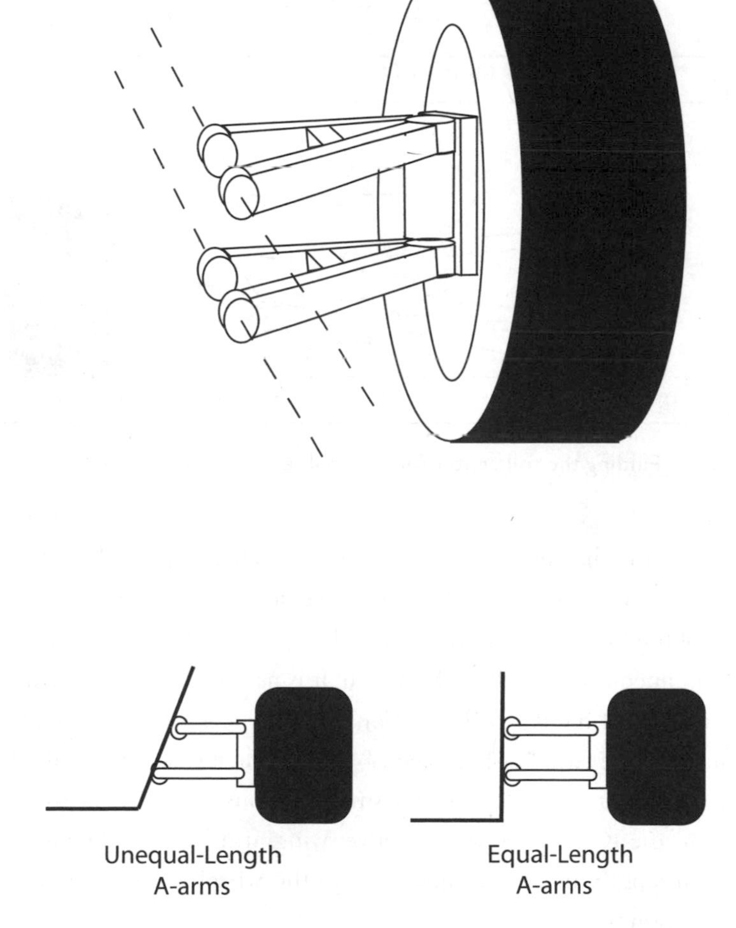

Figure 5.10 Double A-arm suspension.

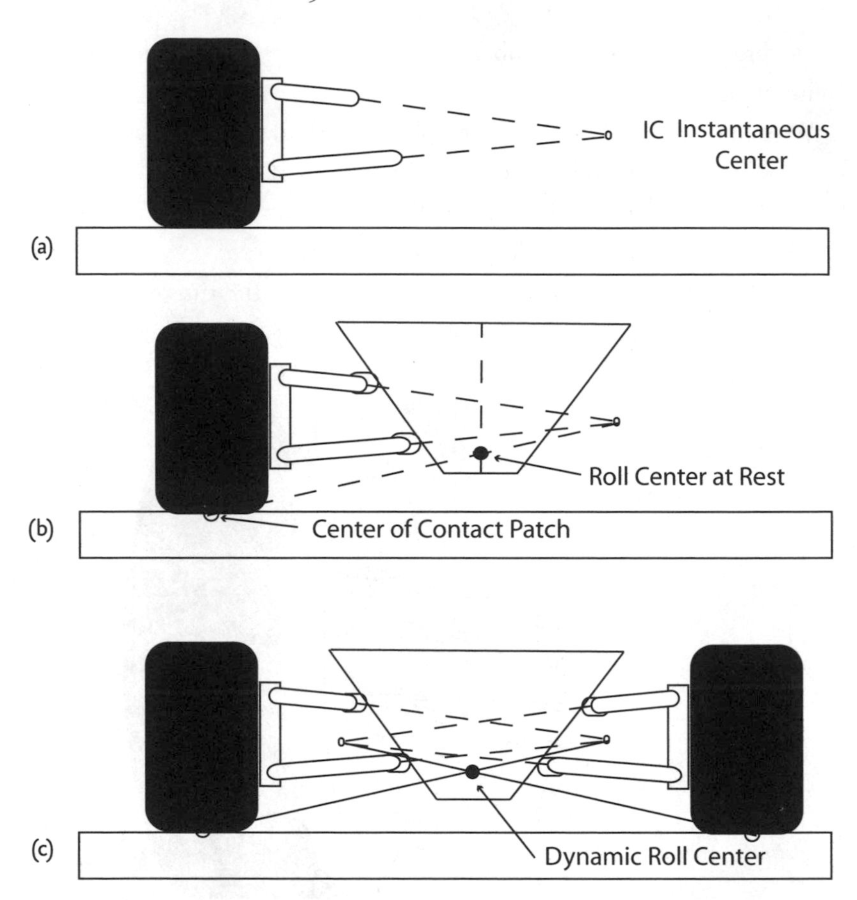

Figure 5.11 Finding the roll center for the double A-arm suspension.

fied chassis, and the upper and lower A-arm links that connect the wheel and chassis. We can visualize the path of the wheel for small motions by projecting the line of the two links until they cross. The point where they cross is called the instantaneous center (IC) of rotation. It is as if the wheel is moving in a circle centered at the IC. If the motions of the wheel are large, the crossing point moves. The term "instantaneous" tells us that it is the center of rotation for the given initial conditions and small motions. The distance from the knuckle to the IC is called the effective swing arm length. The greater the swing arm length, the less camber change the wheel will experience when the suspension moves.

It is interesting that any point on the wheel and knuckle must follow a circular path centered at the IC. We could draw a line from any point on the wheel to the IC and track the circular motion of that point for small deflections. The most important line of this type goes from the center of the contact patch for the tire to the IC. This line behaves like a third link starting at the contact patch and terminating somewhere along the line. If the suspensions on the left and right sides of the car are symmetric, the third imaginary link terminates along the midpoint of the chassis. This point is the instantaneous roll center of the chassis. If forces cause the chassis to roll, it will pivot about this point. As we saw in the case of the instantaneous center for the left wheel, the roll center for the chassis will move with large rotations of the chassis.

If we perform the same analysis for a symmetrical suspension using the right wheel, it will cross the centerline at the same place. This is shown in figure 5.11c. If the suspension is asymmetrical or the roll angle is large, the roll center leaves the centerline of the chassis. To find the roll center in this condition, we find the crossing point for our left and right side suspension lines from instantaneous centers to contact patches. This is shown in figure 5.12a. The height of the roll center relative to the ground is determined by the angle and locations of the control links. Figure 5.12b shows a configuration where the roll center is below ground level.

The front and rear suspensions typically have different roll center heights, with the front typically being lower. The lower roll center will produce flatter motions of the chassis and less camber change in the wheels that do the steering. If we draw a line from the rear suspension roll center to the front roll center, we have identified the roll axis for the car. Figure 5.13 is a side view of the car showing the roll axis and the center of gravity.

In the reference frame of the car under cornering conditions, the inertial force, $m\mathbf{A}$, acts at the center of gravity. The inertial force creates a torque about the roll axis, causing the chassis to roll. Springs and antiroll bars must counter the roll. We can minimize the torque by raising the roll axis closer to the center of gravity. This produces an unfortunate consequence. In figure 5.14 we have simplified our model of the left tire and suspension as just a tire and rigid swing arm. When turning to the right, the friction force at the contact patch points to the right. The friction force produces a torque about the chassis roll center. If the roll center is above the ground, the torque is counterclockwise and the IC rises, pushing upward on the chassis. This push is called a jacking force. Anyone who has seen one of the original VW Beetles under hard cornering has seen the rear end of the car hop up into the air. In figure 5.14b the roll center is below the ground and the torque is clockwise. The resulting jacking force pushes the chassis downward, collapsing the suspension. This

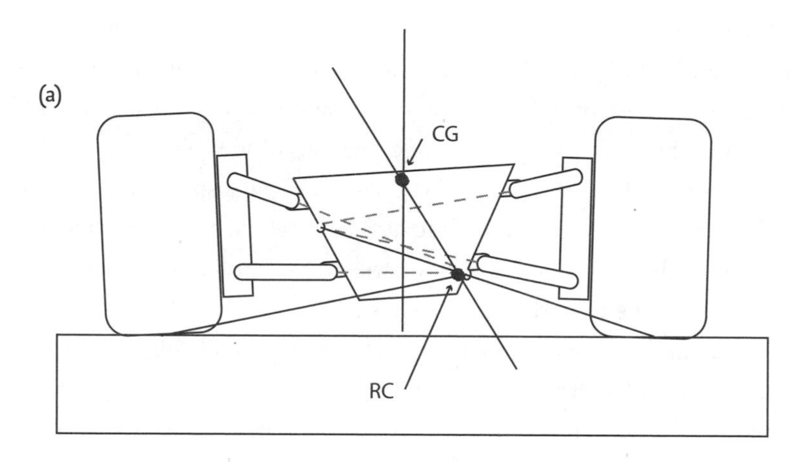

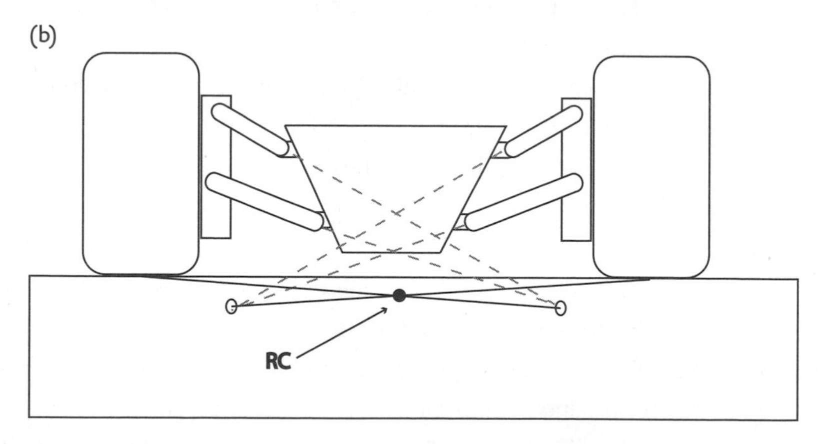

Figure 5.12 (a) Motion of the roll center. (b) The roll center can be below the surface of the road.

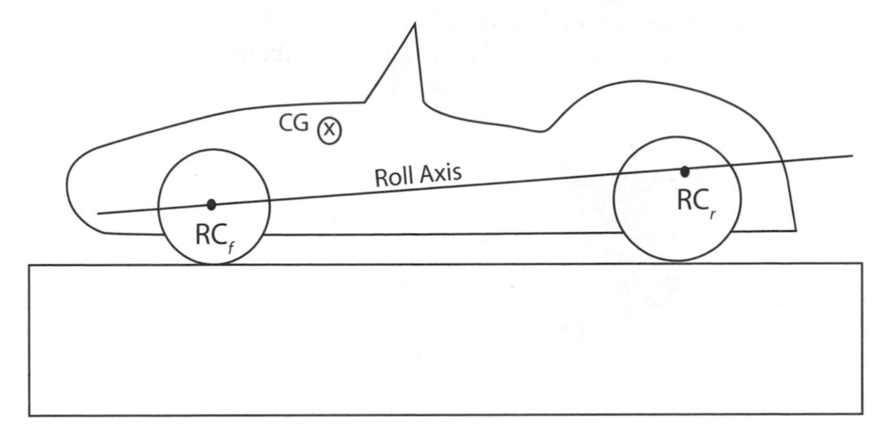

Figure 5.13 Finding the roll axis.

combination of downward jacking force and vertical load transfer can bottom out the suspension, producing abrupt changes in handling and a loss of control. If the roll center is even with the surface of the road, the jacking force is zero.

This leads us to another classic backyard tuner mistake: lowering your car using cut-down springs to lower the center of gravity. This reduces vertical load transfer, and that is a good thing. Lowering an economy car typically gets good results. But what happens when we lower a sports car whose roll center is already at the surface of the road? If the roll center drops below the road surface, we add a downward jacking force. Shorter springs already take away suspension travel. The downward jacking force takes away more travel, and our handling can actually get worse with lowering. We must accept a compromise between load transfer and roll angle.

Using long effective swing arms to minimize camber change with a roll center a few inches above ground level puts us in the ballpark of optimum stability. A shorter upper control arm will buy us some increase in negative camber when the chassis rolls.

Most purpose-built race cars and many sports cars opt for the double A-arm or double wishbone suspension. Open-wheeled racers take it one step further. Suspension components out in the open create aerodynamic drag. To mini-

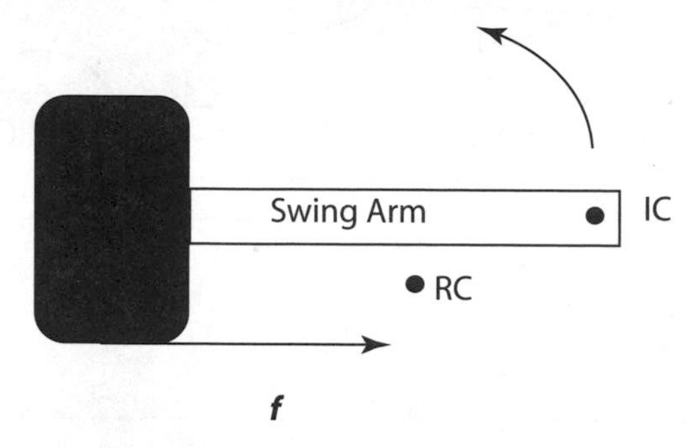

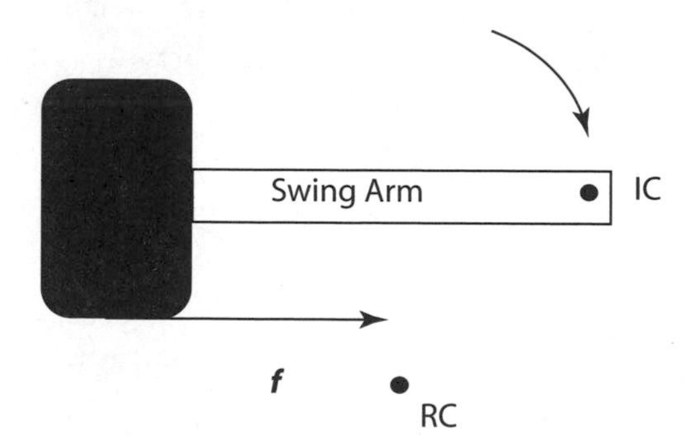

Figure 5.14 Jacking forces for a tire and swing arm.

mize this drag, suspension components are moved inboard and underneath the bodywork. A thin pushrod replaces the spring and shock within the airstream. At the end of the pushrod, underneath the bodywork, is a rocker as shown in figure 5.15. With careful design work, the spring and shocks can be laid horizontally. In some cases, the Ferrari Formula 1 team has replaced the rocker pivot with a torsion bar–type spring and eliminated the need for the coil spring altogether.

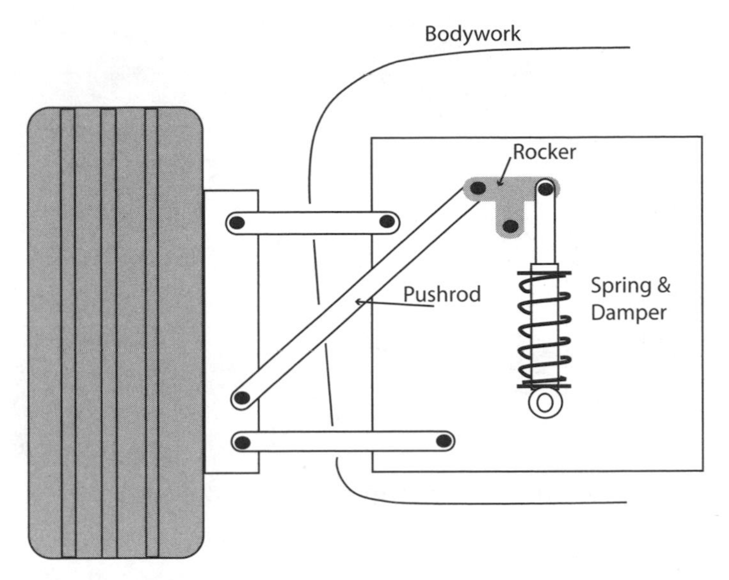

Figure 5.15 A pushrod suspension minimizes aerodynamic drag by moving shocks and springs inside of the body work.

5.6 MACPHERSON STRUT SUSPENSION

The MacPherson strut suspension in figure 5.16 is simple in construction, is relatively lightweight, and works well with front-wheel drive. It is a tall suspension because the entire shock and spring assembly is in the strut located above the wheel knuckle. The strut itself is the upper control arm. Manufacturers use a wide variety of lower control arm assemblies, but it must remain clear of the drive axle. One of the most attractive aspects of the MacPherson strut for the racing crowd is the potential to easily change camber and caster. The top of the strut is bolted to the top of the wheel well. The strut is used primarily in front suspensions, so these bolts are easily accessible under the hood of the car. The fixed bolt assembly is replaced with two sliding plates. One plate is bolted to the chassis, and the other plate is bolted to the strut and can be moved laterally to adjust camber and longitudinally to adjust caster. The two plates are connected by a set of locking bolts to fix the desired alignment parameters.

To find the instantaneous roll center, we project a straight line along the

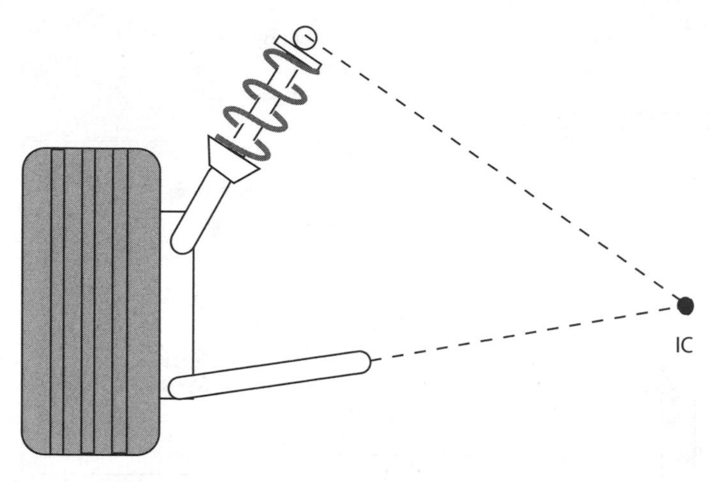

Figure 5.16 MacPherson strut front suspension.

lower control arm as before. We draw the second line perpendicular to the strut passing through the top mount of the strut (see fig. 5.16).

Race cars that are built from the ground up shy away from this design, with the possible exception of off-road rally cars. In fact, the STi employs MacPherson struts in the front. The struts tend to interfere with wide wheels used in road racing. Going to wider tires means pushing the wheel offset outward, and this increases the scrub radius. As we learned previously, this generates unwanted torques about the steering axis, reduces handling responsiveness, and can make handling less predictable to the driver. This system also requires a tall suspension. This works for street sedans but not for low-profile sports cars.

5.7 NASCAR-TYPE SOLID REAR AXLE

In the world of racing and sports cars, most manufacturers and race series have opted for the superior performance of the independently sprung rear end. An independent rear end allows the two rear tires to respond to the road one wheel at a time. In a rear-wheel-drive solid rear axle car, also called a live axle, the wheels are connected to a rigid tube and differential as shown in figure 5.17. Both tires will have the same orientation with respect to the road. Nevertheless, the solid rear axle has hung around for a long time. It is strong, rigid, and relatively simple. Trucks and a few muscle cars like the Ford Mustang continue to employ it on the road. A smooth track mitigates some of the advantages of an independent rear end. Whether the reason is historical, economical, or technological, NASCAR has stuck with the solid rear axle. Since NASCAR is the most popular motorsport in the United States, we should understand the effects of the solid axle.

Figure 5.17 is a particular type of solid axle suspension called the truck arm solid axle. Motion is constrained by two trailing arms and a Panhard bar. The trailing arms are I-beams bolted to the axle with large U-brackets in a fixed orientation. As we trace the beams forward from the axle, they are parallel to the bottom of the chassis and are canted inward toward the chassis centerline. If we project the beams forward, they appear to cross at a single point, which forms an instantaneous center (IC) of rotation. The beams never reach this point. Instead, they cut off on either side of the driveshaft and are connected

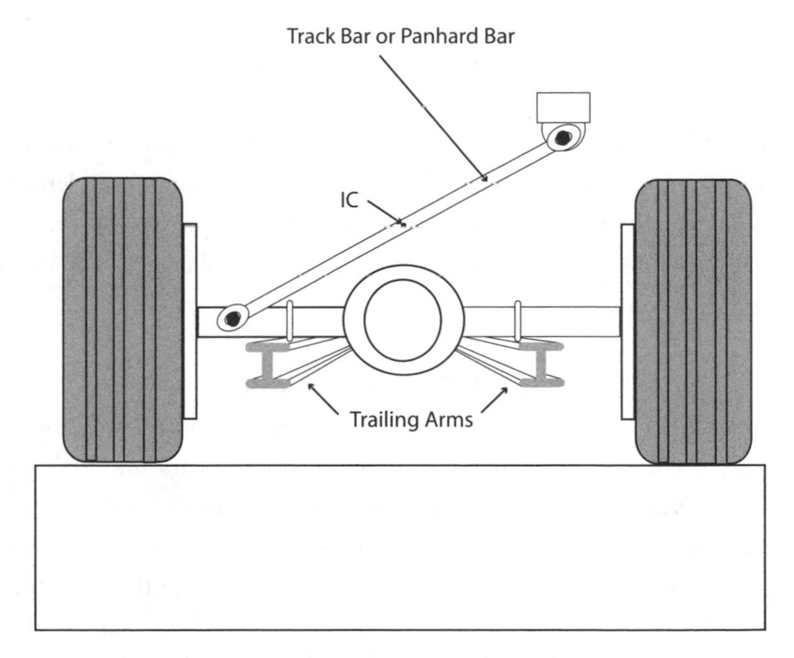

Figure 5.17 NASCAR-style solid rear axle with trailing truck arms and track bar.

to the chassis by a pivot that is oriented lateral to the car. The beams allow the rear axle to move up and down freely. The beams are long and allow the rear axle to flex laterally, both left and right with respect to the car. This motion is constrained by the Panhard bar, also called a track bar. The track bar is a long metal rod with pivots at either end. The pivots are oriented longitudinally with respect to the car. One end is connected to the axle housing on the left of the car, and the other end is connected to the chassis on the right side of the car. For small roll angles of the chassis, the midpoint of the track bar forms a second IC. The roll axis for the rear of the car is the line joining this IC with the IC just forward of the beams. This IC for the rear suspension is on this line just above the differential as shown in figure 5.17.

The track bar upper pivot point where it connects to the chassis has an adjustable height. Changing the track bar height changes the rear roll center height of the car. Changing the roll center height changes the vertical load transferred at the wheels and changes the handling of the car. We'll come back to this after we discuss springs, but it is important to know that this is one of the adjustments that can be accomplished during a pit stop. A mechanic lowers a ratchet with a long extension through a hole in the right side of the back window. It engages a bolt that adjusts the height of the attachment point for the track bar.

For left-hand turns, the track bar is under tension. This is an inherently stable condition. This favors handling on ovals that contain only left-hand turns. For right-hand turns, the track bar is under compression, a somewhat less stable condition. Maybe this explains part of the challenge that NASCAR drivers face when they take on a road course.

5.8 SPRINGS AND DAMPERS

We control the motion of the suspension with springs and dampers. Springs provide a restoring force, which pushes the suspension back toward its equilibrium position where the net force is zero. Springs store energy when they are stretched or compressed. This stored energy is converted to kinetic energy, or energy of motion, as the suspension passes back through the equilibrium position. The result is an overshoot of equilibrium, and the energy is passed back into the spring. This ideal process of oscillation is endless until we install

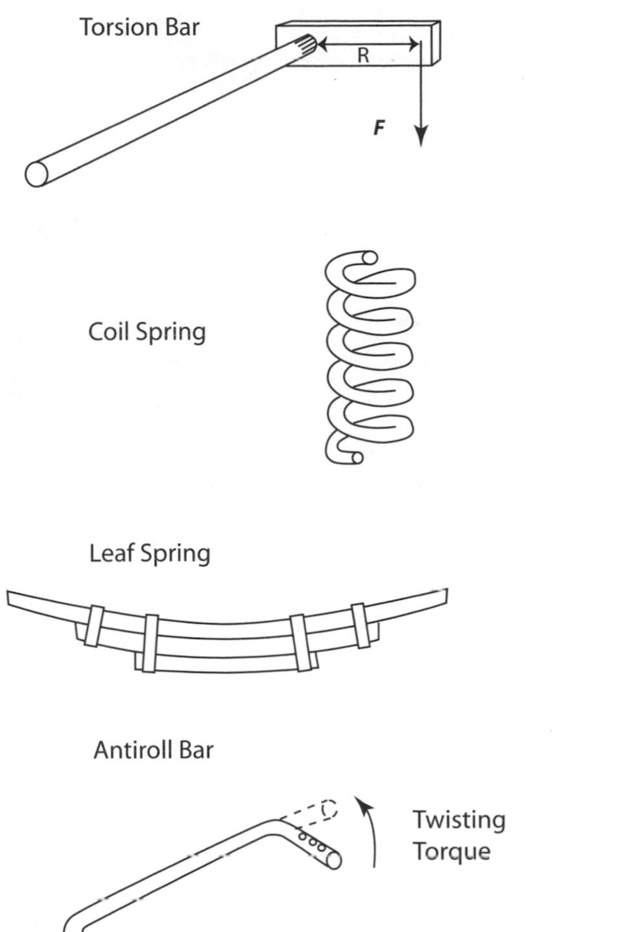

Figure 5.18 Various spring types.

a damper or a shock absorber. The damper converts the mechanical energy into thermal energy and suppresses the motion.

Sports cars employ a variety of spring types, including torsion bars, coil springs, and leaf springs, shown in figure 5.18. In general, they rely on shear stress in the spring metal. Shear stress is a force, F, per unit area, A, applied tangent to a surface as shown in figure 5.19a. The material deforms a distance Δx. That distortion is spread over the thickness, L, of the material. We define the shear modulus, S, as the shear stress, F/A, divided by the shear strain, $\Delta x/L$:

$$S \equiv \frac{F/A}{\Delta x/L}.$$

The shear modulus is a property of the material being considered. It also depends on how the material is processed, such as the heat treatment or annealing. A typical value of 10 to 12 million pounds per square inch is employed in iron and steel torsion bars.

We must slightly rethink the shear geometry to come up with a spring constant for a torsion bar. Our plan is to chop the solid cylindrical torsion bar up into a series of hollow cylinders and estimate its spring contribution. We will then add up the contributions from each hollow cylinder. Figure 5.19b shows a

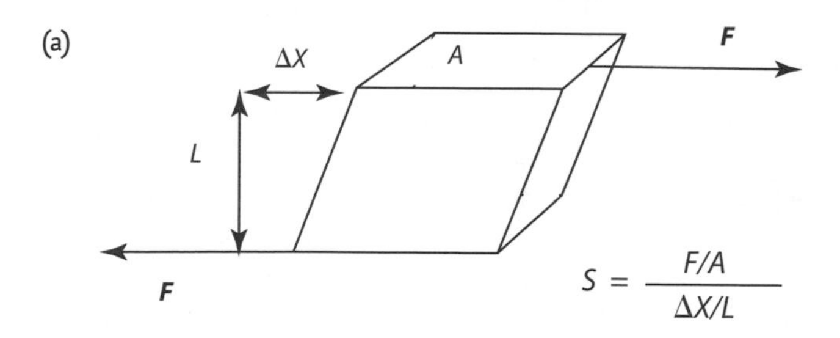

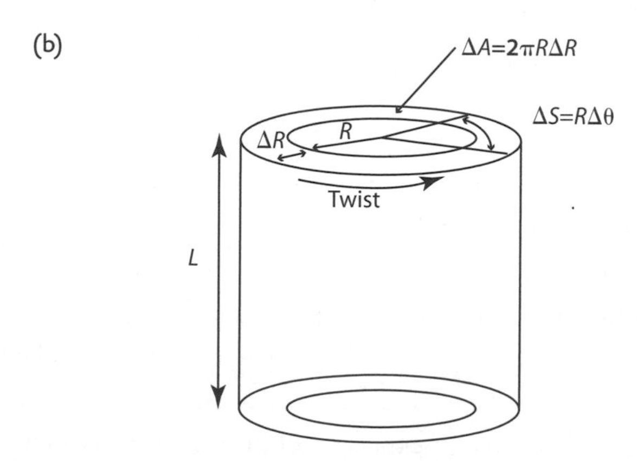

Figure 5.19 Shear stress and the torsional spring.

thin hollow cylinder that is twisted around the cylindrical axis so that the shear force, ΔF, is tangent to the surface of the cylinder. The delta indicates that it is a small contribution that we will need to add to the other hollow cylinders. The force is applied a distance R from the center of the cylinder, creating a torque, $\Delta\tau = R\Delta F$. If we can figure out the proper expressions to plug into our shear modulus equation, then we can find an expression for the force. Here Δx is a piece of arc length $R\Delta\theta$. The area is $\Delta A = (2\pi R)\Delta R$. L is the length of the cylinder.

Rearranging the shear modulus equation and substituting in our new expressions, we have

$$\begin{aligned}\Delta\tau &= R\Delta F \\ &= R\frac{R\Delta\theta}{L}(2\pi R)\Delta RS \\ &= \frac{2\pi S}{L}R^3\Delta R\Delta\theta.\end{aligned}$$

If you have taken some calculus, you may recognize that the deltas can be replaced with differentials and we can integrate over all of the hollow cylinders and solve the equation:

$$\tau = \frac{S\pi}{32L}\theta d^4 = k\theta.$$

Here d is the diameter of the torsion bar, θ is the angle of the twist in radians, L is the length of the torsion bar, and S is the shear modulus of the material. Our equivalent spring constant is k. The fourth power dependence on the diameter is important. It leads us to the idea that the outermost material is the most important. It experiences the greatest distortion. Antiroll bars are a type of torsional spring. For a long time designers have used hollow cylindrical tubes, since most of the spring constant comes from the outermost material. Hollowing out the torsion springs significantly reduces their weight. Developing a similar expression for coil springs is more challenging. The result is relatively simple and useful:

$$F = \frac{Sd^4x}{8D^3N} = kx.$$

Here d is the diameter of the wire, D is the diameter of the coil loops, N is the number of loops, and x is the distance that the coil is compressed. The actual spring rate depends on the application. Lightweight race cars, without aerodynamic down-force, may run springs with 200 lbs/in. NASCAR mandates 500 lb/inch springs at the rear. A high down-force, high-speed racer may double that number.

Some aftermarket suspension components use a small "helper" spring on top of the main spring. What is the spring constant for this combination? The key to answering this is recognizing that both springs experience the same force, F. This gives us the following equations for spring constants k_1 and k_2:

$$x_1 = \frac{F}{k_1} \text{ and } x_2 = \frac{F}{k_2}.$$

Spring 1 compresses a distance x_1 and spring 2 compresses a distance x_2. The total compression is $x_T = x_1 + x_2$. The total effective constant $k_T = F / x_T$. Putting these equations together,

$$x_T = x_1 + x_2$$

$$\frac{F}{k_T} = \frac{F}{k_1} + \frac{F}{k_2}$$

$$k_T = \frac{k_1 k_2}{k_1 + k_2}.$$

The total effective spring constant is less than that for either spring alone. This is an interesting result. Physicists love analogous systems. This has a direct analogy in electrical capacitors, which, like springs, store energy. We could calculate the effective capacitance of capacitors in series, as we did with our springs, by adding the inverses. One difference for the spring system is that once the helper spring is completely compressed, we are left with just the spring constant of the main spring.

If two coil springs of equal length are placed concentrically with each other and compressed by a single force, the effective spring constant k_T is

$$k_T = k_1 + k_2 .$$

This is the same form that we have for the equivalent capacitance of capacitors connected in parallel. While we won't consider concentric coil springs, we will tackle the case in which the antiroll bars act in parallel with coil springs. An antiroll bar, also called a sway bar, is shown is figure 5.18. It is a U-shaped horizontal bar where the base of the U is connected to the chassis. The chassis connection is made via bushings that allow the base to rotate freely. The two arms of the U are connected to the lower control arms of the left and right suspensions. If you drive over a speed bump, the left and right tires push the suspension upward at the same time. Both arms of the antiroll bar move upward with the suspension and the bar does nothing. However, if only one tire hits a bump on the right side, the suspension and the antiroll bar lever arm on the right side move upward. The arms of the U shape are twisted out of the same plane and the arm on the left side tries to compress the spring on the left. If the antiroll bar were infinitely stiff, the left and right side coil springs would be forced to compress equally for a road bump on only the right side. The overall effect would be to double the spring rate with left and right coil springs acting in parallel. In a turn the chassis would be prevented from rolling to the outside, since the inside and outside springs must compress equally. A small degree of roll would still exist because the tires still act like uncoupled springs. A large fraction of the vertical load is quickly transferred to the outside tire. Of course, antiroll bars aren't infinitely stiff. When the car rounds a corner, the antiroll bar twists, transfers some vertical load to the outside, and resists but does not prevent chassis roll. We'll come back to the antiroll bar at the end of the chapter.

Now that we know a little about springs, it is time to begin to model the suspension. As is typical in physics, we will start with a simplified model and move toward the more complex reality. Figure 5.20 shows a simple system with a mass connected to a spring. It is called a simple harmonic oscillator and is one of the fundamental models in physics. It is used to develop an understanding of electrical circuits, vibrating atoms, and automotive suspensions. It is worth a little time to review some basics before we turn directly to automotive suspensions.

There is a vertical coordinate system with $z = 0$ at the equilibrium position. As we have seen, springs typically produce a force that is proportional to the displacement from the equilibrium position. This is Hooke's law:

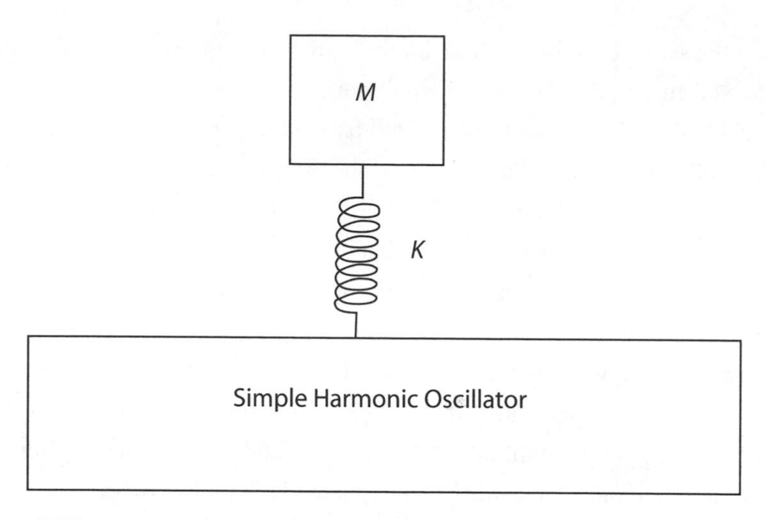

Figure 5.20 A mass and spring that constitute a simple harmonic oscillator.

$$F = -kz. \tag{5.1}$$

F is the force in pounds, z is the position with respect to the unloaded equilibrium position in inches, and k is the spring constant in pounds per inch (or N/mm). Applying Newton's second law,

$$\begin{aligned} \sum F_z &= ma_z \\ -kz &= ma_z. \end{aligned} \tag{5.2}$$

This is a differential equation. If you are aren't familiar with calculus or differential equations, the next couple of steps will be foreign, but useable results will quickly follow. Velocity in the z-direction is the rate of change of the position. Acceleration in the z-direction is equal to the time rate of change of velocity in the z-direction:

$$V_z = \frac{dz}{dt} \tag{5.3}$$

$$a_z = \frac{dV_z}{dt} = \frac{d^2z}{dt^2}, \text{ so} \tag{5.4}$$

$$-kz = m\frac{d^2z}{dt^2}. \tag{5.5}$$

This means that z is some function of time that, when differentiated twice, gives the original function times a negative constant. This means that z is a sine or cosine function, or an exponential. For simplicity, let's use a generalized cosine function:

$$z = A\cos(\omega_0 t + \phi) \tag{5.6}$$

$$\omega_0 = \sqrt{\frac{k}{m}} = \text{natural angular frequency} \tag{5.7}$$

$$\phi = \text{phase constant.} \tag{5.8}$$

A is the amplitude of the oscillation or the maximum distance that the mass oscillates above or below the equilibrium position in units of inches, t is the time in seconds, ϕ is the phase constant that determines where in the oscillation cycle the mass is located at time $t = 0$, and ω_0 is the natural frequency in radians per second. If we divide it by 2π, we get the natural frequency, f_0, in cycles per second or hertz. The period of oscillation, T, is the time it takes to go through one full cycle of oscillation. The period is equal to the inverse of the frequency, $T = 1/f_0$. Figure 5.21, a plot of z as a function of time, is labeled

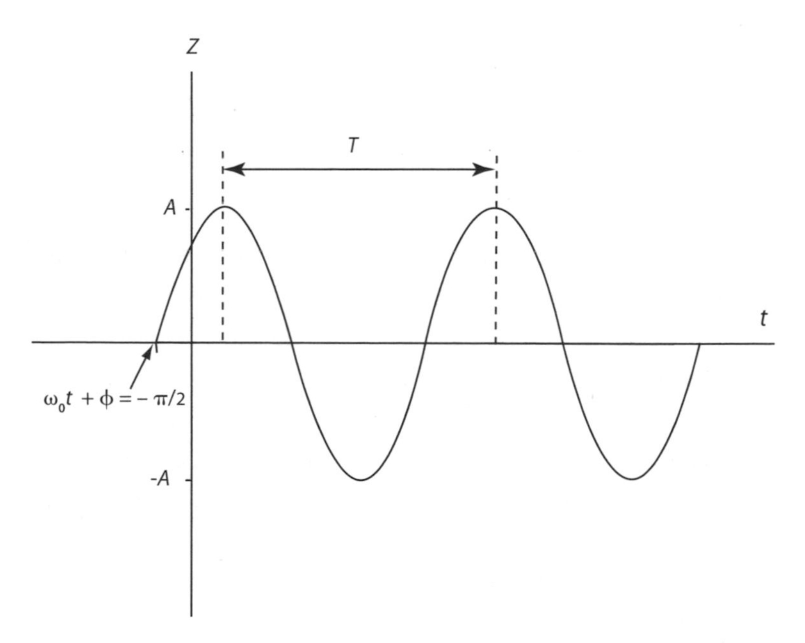

Figure 5.21 General cosine function with amplitude A, period T, and phase constant, ϕ.

with the various parameters. We can find the velocity and acceleration of the mass using equations (5.3), (5.4), and (5.6):

$$V = \frac{dz}{dt} = -A\omega_0 \sin(\omega_0 t + \phi) \tag{5.9}$$

$$a = \frac{dV}{dt} = -A\omega_0^2 \cos(\omega_0 t + \phi). \tag{5.10}$$

$A\omega_0$ is maximum speed of the mass, and $A\omega_0^{\ 2}$ is the maximum acceleration.

The kinetic energy, K, the energy associated with the motion of the mass, is equal to $\frac{1}{2}mV^2$. The potential energy, U, the energy stored in the spring, is $\frac{1}{2}kz^2$. The total mechanical energy of the system, E, is the sum of the kinetic and potential energies. Substituting equations (5.9) and (5.10), we obtain

$$K = \frac{1}{2}mV^2 = \frac{1}{2}mA^2\omega_0^2 \sin^2(\omega_0 t + \phi)$$

$$U = \frac{1}{2}kz^2 = \frac{1}{2}kA^2 \cos^2(\omega_0 t + \phi) \text{ and } \omega_0^2 = \frac{k}{m}$$

$$U = \frac{1}{2}mA^2\omega_0^2 \cos^2(\omega_0 t + \phi)$$

$$E = K + U = \frac{1}{2}mA^2\omega_0^2 \left[\sin^2(\omega_0 t + \phi) + \cos^2(\omega_0 t + \phi)\right]$$

$$E = \frac{1}{2}mA^2\omega_0^2 = \frac{1}{2}kA^2.$$

Based on the trigonometric identity that $\sin^2(\theta) + \cos^2(\theta) = 1$, we see that the total mechanical energy is a constant. If we plot the potential energy, U, as a function of the position, z, as in figure 5.22, we can add a horizontal line that corresponds to the total mechanical energy. The distance from the z-axis up to the curve corresponds to the potential energy, U. The distance from the curve up to the horizontal line of the total mechanical energy, E, corresponds to the kinetic energy, K. As the mass oscillates from $-A$ to $+A$, the energy sloshes back and forth from potential to kinetic and back to potential energy.

To suppress this motion, we must convert some of the energy to another form. A simple damper would be to suspend a disk in a viscous fluid as shown in figure 5.23.

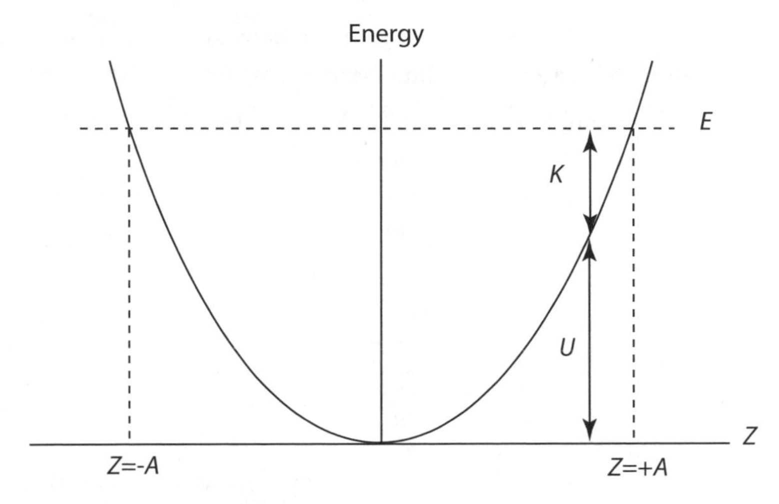

Figure 5.22 Potential energy as a function of position for a simple harmonic oscillator. E is the total mechanical energy, U is the potential energy, and K is the kinetic energy.

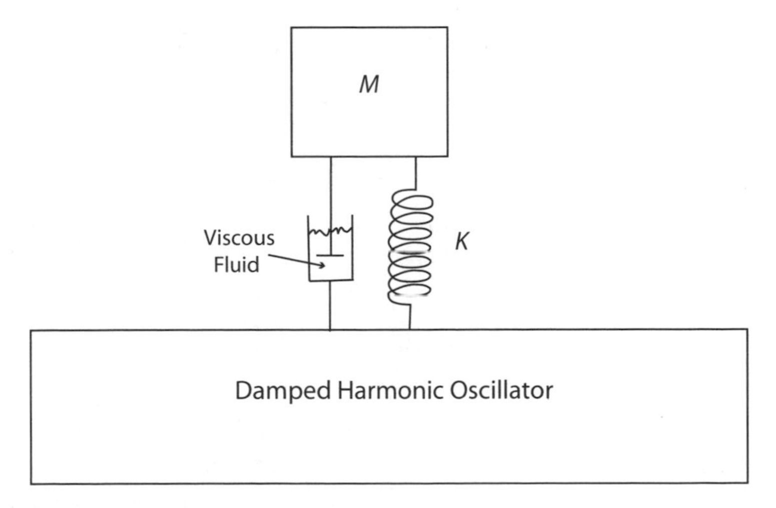

Figure 5.23 A disk in a viscous fluid adds damping to our simple harmonic oscillator.

At low speeds, the damper produces a force that resists the motion of the mass and is proportional to the speed. The mechanical energy of the mass will end up as thermal energy in the damper fluid. If the disk in the damper is small and the viscosity of the fluid is small as it is in water, the mass may bounce many times before most of the energy is converted into a temperature rise in the fluid. We call this condition underdamping. On the other hand, if the disk is large and the fluid is highly viscous like molasses, the damper may barely seem to move as it returns to equilibrium. We call this condition overdamping. We can again use Newton's second law as a starting place to develop a mathematical description of the motion. Applying Newton's second law with a little calculus,

$$-kz - Bv = ma_z \tag{5.11}$$

$$m\frac{d^2z}{dt^2} + B\frac{dz}{dt} + kz = 0. \tag{5.12}$$

B is a constant associated with the damping, m is the mass, and k is the spring constant. The simplest form for the underdamped solution is a cosine oscillation with amplitude (shown in the square brackets) that decays exponentially:

$$z = \left[A_0 e^{-(B/2m)t}\right]\cos(\omega' t + \phi) \tag{5.13}$$

$$\omega' = \omega_0\sqrt{1 - \left(\frac{B}{2m\omega_0}\right)^2}. \tag{5.14}$$

For simplicity, let's assume that the phase constant is zero. This just means that at time $t = 0$ the amplitude is at a maximum. Figure 5.24 shows the three general types of behavior observed in this system.

The system oscillates endlessly without damping. When we add damping to the system, the angular frequency decreases and the amplitude decreases with each oscillation. This is the underdamping case. When the damping coefficient, B, is equal to $2m\omega_0$, ω' becomes zero and oscillation stops. We are left with a decaying exponential that returns to equilibrium in the least amount of time without overshooting equilibrium. We call this case critical damping and label the damping coefficient as B_C. In the overdamping case, the damping is greater than B_C and the mass takes longer to return to equilibrium. You might think that automotive engineers opt for critical damping. In fact, they typically choose a slightly underdamped system. In this instance, the

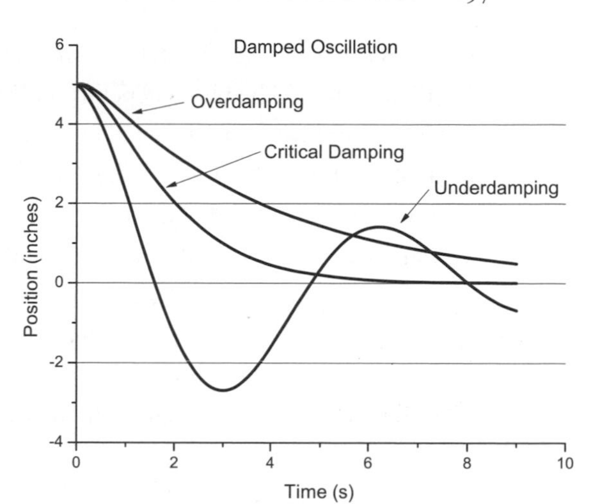

Figure 5.24 Damped oscillation as a function of time.

mass will more quickly pass through equilibrium followed by small-amplitude oscillations. In this way, the system is quickly ready for the next bump.

The damped oscillation, which we have just considered, is a decent description of a single-step bump. What happens when the road is a washboard and the vibrations happen at a fixed interval? This is an example of a driven oscillation. The simplest mathematical driving force is one that varies as $\cos(\omega t)$. Newton's second law can be written as

$$m\frac{d^2z}{dt^2} + B\frac{dz}{dt} + m\omega_0^2 z = F_0\cos(\omega t) \tag{5.15}$$

$$\omega_0 = \sqrt{\frac{k}{m}}\,. \tag{5.16}$$

F_0 is the amplitude of the driving force and ω is the driving angular frequency of the washboard road. An equilibrium solution is another cosine function:

$$Z = A\cos(\omega t - \delta) \tag{5.17}$$

$$A = \frac{F}{\sqrt{m^2(\omega_0^2 - \omega^2) + B^2\omega^2}} \tag{5.18}$$

$$\tan\delta = \frac{B\omega}{m(\omega_0^2 - \omega^2)}\,. \tag{5.19}$$

The cosine function has a couple of new twists. First, the cosine is out of phase with the driving force by an amount δ. The phase depends on the mass, the damping, the driving frequency, and the natural frequency. The second feature is a frequency-dependent amplitude. Figure 5.25 shows the amplitude as a function of driving frequency for three different damping constants.

The sharp peak is a resonance peak centered at the resonant frequency, ω_0. If we drive the mass spring system at the right, frequency of the amplitude grows. The less damping we have, the greater the peak. This kind of behavior is typical of all oscillating systems. It is the job of the designer or tuner to set the damping and the resonant frequency to control the oscillation of the mass.

In automotive engineering, our mass-spring-damper system is called a quarter-car model because it neglects interactions with the other three corners of the car. We can improve the accuracy of our quarter-car model by considering the effect of the elastic properties of the tire. We have already seen that the tire displays both viscous and elastic properties. This is modeled as an additional spring and damper as shown in figure 5.26.

The unsprung mass includes the tire, wheel, brake, hub, and knuckle

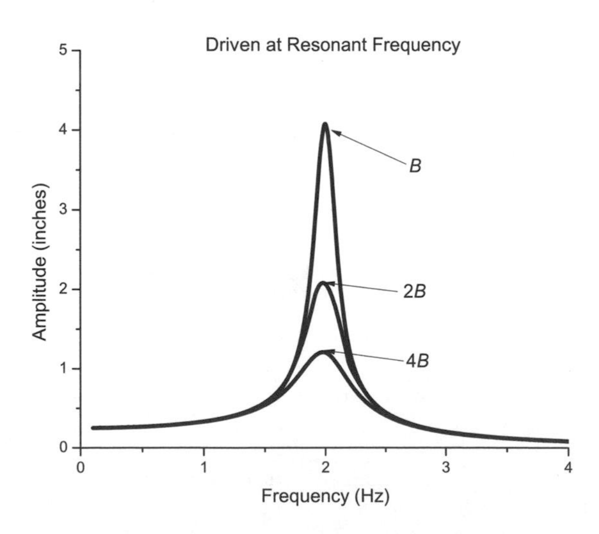

Figure 5.25 Damping amplitude as a function of driving frequency and damping constant *B*.

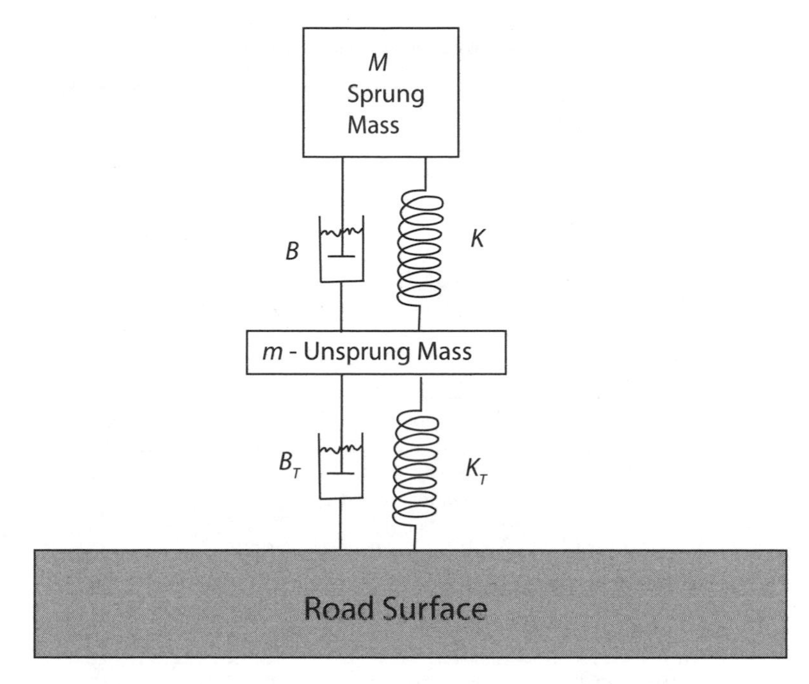

Figure 5.26 Harmonic oscillator with both a sprung and unsprung mass.

and a portion of the axle and suspension components. It is in series between the spring/damper formed by the tire sidewall and the spring/damper that connects the suspension to the car chassis. The detailed solution is beyond the scope of this book. Richard Stone and Jeffery Ball extend the solution a little further in *Automotive Engineering Fundamentals*. William Milliken and Douglas Milliken also consider some of the details in *Race Car Vehicle Dynamics*. We can qualitatively consider some of the details by considering figure 5.27, which shows body acceleration as a function of frequency when our model is expanded to incorporate the tire spring constant and the unsprung mass.

We now have two peaks. The first, at around 2 Hz, is from the resonant oscillation of the sprung mass (the body) on the springs. The second is the resonant oscillation of the unsprung mass between the spring and the flexing tire. The tire spring constant, k_T, and tire damping constant, B_T, are functions of the tire we select. If we minimize the unsprung mass and opt for stiff

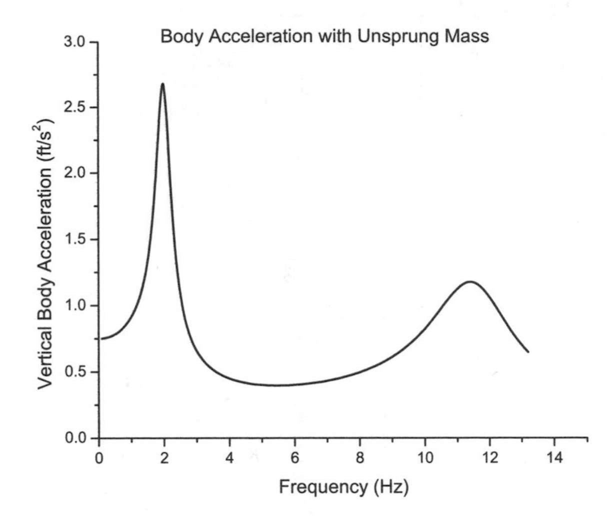

Figure 5.27 Vertical body acceleration as a function of frequency. Note the two different resonant frequencies.

sidewall tires, we can raise the frequency of the second peak and minimize its effect on handling. Of course, the stiff sidewall will transmit high-frequency road noise to the chassis, a negative if you are building a luxury car. Minimizing the unsprung weight is another problem area for aftermarket tuners. Two typical upgrades, bigger wheels and bigger brakes, add to the unsprung weight and lower the resonant frequency of the second peak. Big oscillations of the unsprung weight can cause the tires to lose traction at best and lose contact with the road in the worst case.

5.9 SHOCK ABSORBERS

Shock absorbers or dampers come in a variety of designs. Figure 5.28 is a generic sketch of a monotube type and exhibits many of the common features.

The upper shaft is connected to the chassis at the top and to a piston at the bottom. Hydraulic fluid resides in the cylinder both above and below the piston. Fluid passes through the piston via an orifice or a disc valve. The characteristics of the fluid and these passages determine the damping constants. The damper converts mechanical energy of the chassis and unsprung weight into thermal energy. The temperature of the hydraulic oil rises. Pressure changes drastically, especially as the oil flows through the passages. In the lowest pressure areas the oil can cavitate or briefly flash to a vapor. One way to prevent cavitation and boiling of the oil is to increase the static pressure. In the monotube damper shown in figure 5.28 a second floating piston is added with hydraulic oil above it and pressurized gas below it. The gas is typically nitrogen pressurized from 100 to 300 psig. A bump in the road forces the piston rod into the damper body, and the piston rod displaces some of the oil. Since the oil is nearly incompressible, it presses down on the floating piston and compresses the gas. This action increases the effective spring rate of the damper and spring combination. Spring rate is an automotive term for the spring constant. We will use the terms interchangeably.

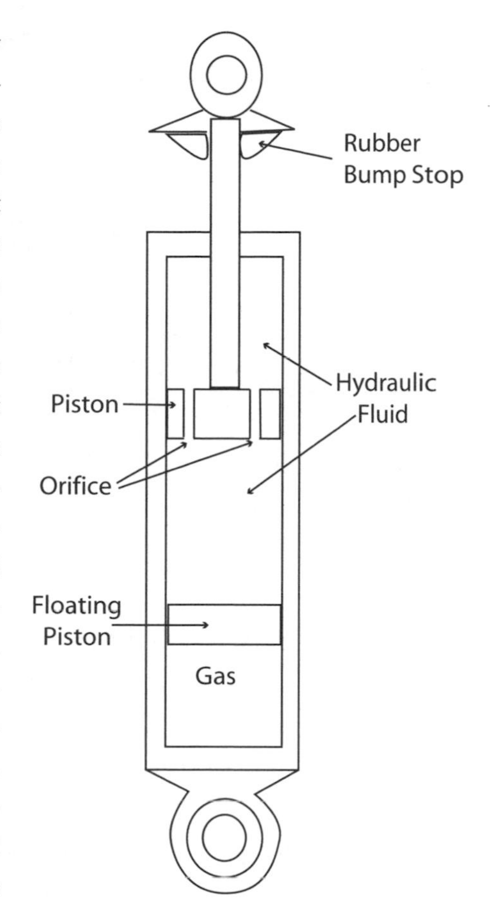

Figure 5.28 Automotive monotube damper, also called a shock absorber.

Once the spring rates and tires are chosen and the sprung mass is set, the resonant frequencies are fixed. Now we must choose the damping rate. We know it must be less than critical damping to ensure responsiveness. If it is too little, it will fail to control resonance in the unsprung mass. We must consider other factors to nail down the damping. Figure 5.29 shows the force as a function of velocity for the damping as we modeled it to this point, with the magnitude of the force, F, equal to BV. V is the velocity and B is the damping constant, which is equal to the slope of figure 5.29.

When our car runs over a bump in the road, inertia of the chassis resists the

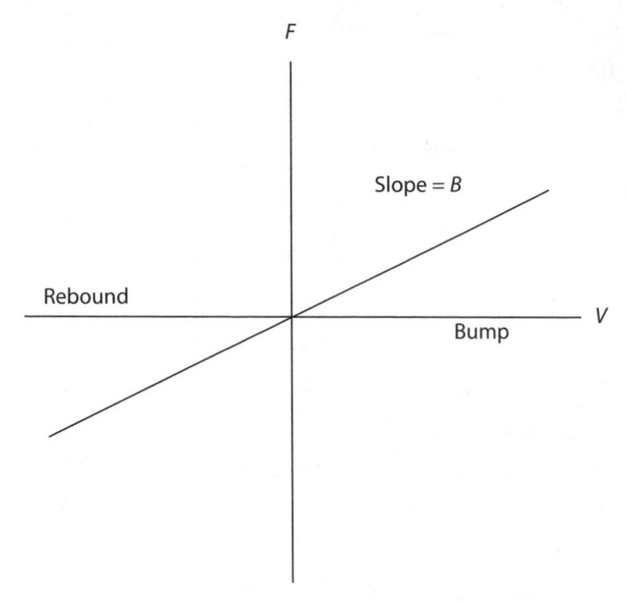

Figure 5.29 Force exerted by a damper as a function of the damper piston velocity.

motion. The unsprung mass must move rapidly to accommodate the compression direction, also called the bump direction. Velocity in the rebound direction is significantly slower and thus the damper must help to control motion of the entire chassis. A significantly greater damping constant is required in the rebound direction. This is shown in figure 5.30a. It is common practice to replot both bump and rebound in the same quadrant as shown in figure 5.30b.

Different frequency regimes correspond to different wheel vertical speeds. It is not surprising that these different regimes can be optimized by different rebound damping constants. Figure 5.31 shows a typical break in the rebound slope.

If we were only considering passenger cars, our work might be done. In racing vehicles, however, the dampers must also control transient load transfer between wheels. The dampers only exert force during the transition. Therefore, the dampers play a role on corner entry and exit. If the dampers control suspension motion properly on corner entry, the suspension will enter equilibrium through the middle of the corner. The car's attitude and load distribution will be determined by the springs, sway bars, and roll center height and not by the dampers. As the car exits the corner, the load distribution is in transition and once again the dampers apply a force and help to control the traction.

Suspension input transients such as road surface bumps and curb hops happen quickly. Load transfer between the wheels due to cornering happens over a much longer period of time. Therefore, we control load transfer using

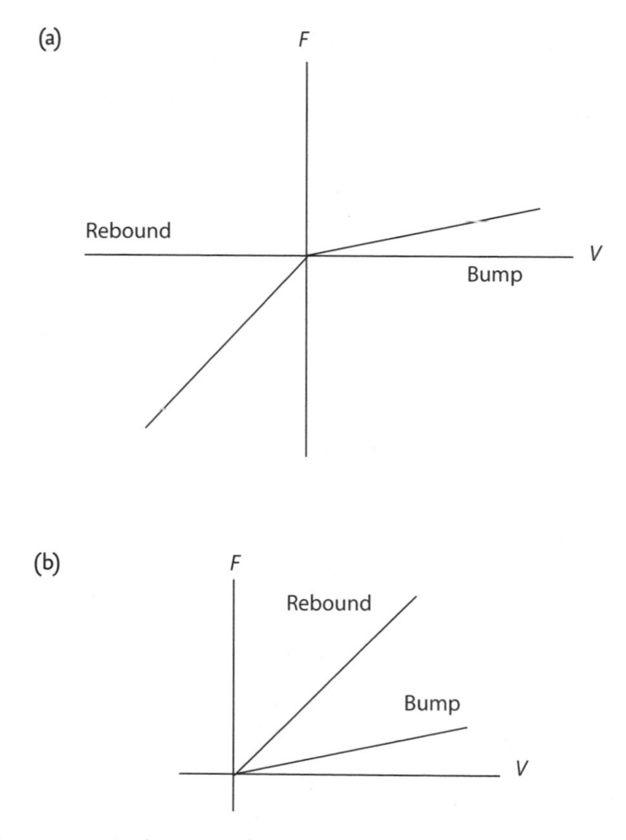

Figure 5.30 Damper force as a function of piston velocity with asymmetric damping. (a) Positive velocities are in the bump direction and negative are in the rebound direction. (b) Both bump and rebound are displayed in the same quadrant.

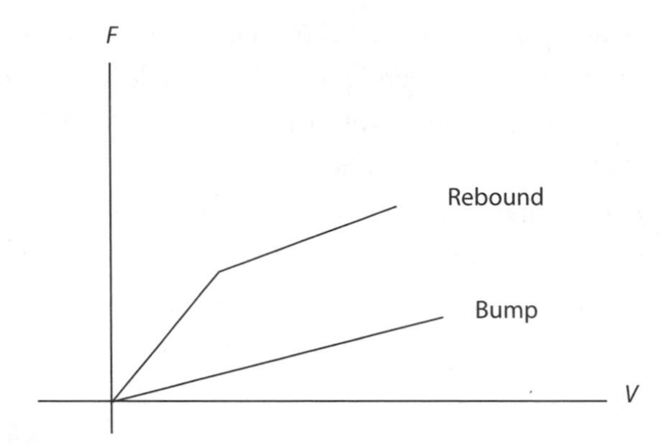

Figure 5.31 Rebound damping with different damping constants for low and high speed.

the low-speed bump and rebound characteristics of the damper. For example, one option for improving corner entry understeer is to increase the low-speed bump front damping. This will slow the rate of body roll to the outside, which can help with steering response. Unfortunately, it will also increase the total load transfer at the front, which will slightly reduce the total front lateral force. Most adjustments involve some type of trade-off. We will return to shock tuning later.

Race teams arrive at the track with an idea of the damping characteristics that they will need. The initial damping information comes from records of the previous year's setup for returning cars or based on current year testing for new cars. These initial settings are rarely the final race setup. If the dampers are allowed to be externally adjustable, the teams can easily begin the dialing in of the settings. For race series that require nonadjustable dampers, the teams must swap out the dampers entirely or rebuild the dampers between practice runs. The bigger teams have a shock dynamometer to produce curves like those in figure 5.31 while they are at the track.

5.10 LATERAL LOAD TRANSFER: ADVANCED APPROACH

In chapter 4 we showed that lateral acceleration causes vertical load to be transferred from the inside wheels to the outside wheels. We can rewrite equation (4.6) in terms of the load transfer, ΔW:

$$\Delta W = \frac{WAh}{T}. \tag{5.20}$$

W is the weight, A is the acceleration in g's, h is the height of the center of gravity above the ground, and T is the wheel track width. Note that Ah/T is dimensionless.

Lateral load transfer has two significant aspects that we have not considered. The first issue is that load transfer reduces the overall traction. Consider a left-hand turn where the load transfers to the outside or right-hand side of the car. The vertical load gained by the right-hand tires is lost by the left-hand tires. If the coefficient of friction were equal on the two sides of the car, the net lateral force would be constant. However, you may recall from chapter 4 that, as the vertical load increases, the effective coefficient of friction for tires decreases. This means that the lateral force gained by the right-hand tires is less than that lost by the left-hand tires. The overall lateral force must decrease. If the vertical load gets sufficiently large on a tire, the lateral force

Controlling load transfer is a key to improved handling. Here Eric Kriemelmeyer maxes out the rear weight transfer in his 2003 Nissan Sentra SpecV. Photograph by Clyde Caplan and Alex Teitelbaum.

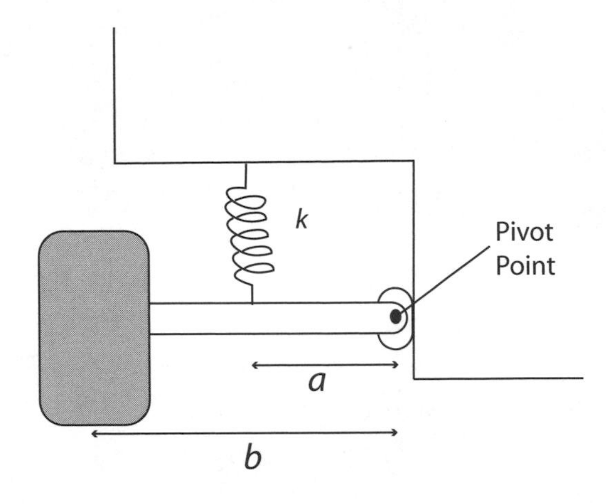

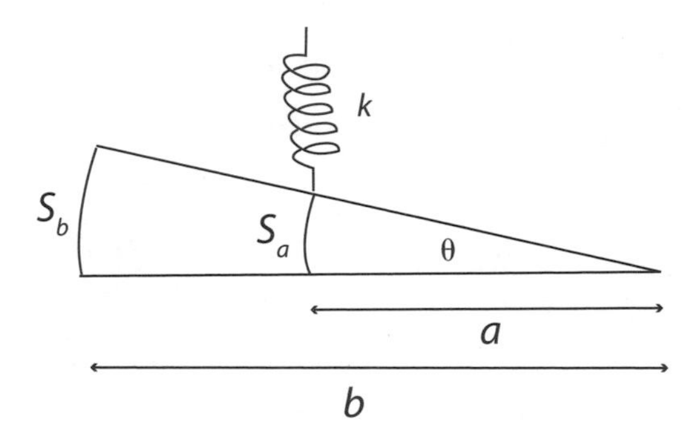

Figure 5.32 Simplified model for determining wheel rates from spring rates.

will pass through a maximum and start to drop (see fig. 4.20). The overall loss of traction will be catastrophic. There are only a couple of ways to reduce the vertical load transfer for a given lateral acceleration. We can lower the center of gravity, h, we can widen the track, T, or we can reduce the weight, W.

The second question is, how is the lateral load transfer distributed between the front and the rear of the car? It is a challenge to resolve the physics, but it is very important from a handling perspective. We can correct understeer or oversteer by controlling the ratio of the front to rear load transfer. We need to come up with a method to estimate the front and rear load transfer. Some choose to approach the problem with a half-car model, considering the roll center and spring characteristics of only half of the car. The problem with this approach is that it neglects the fact that the two ends of the car are coupled. For example, change only the spring characteristics at the front of the car and you change the weight transfer characteristics at the rear of the car. Instead, we will consider the entire car and then divide up the torques based on the relative front and rear spring rates.

We can't use the spring rates directly. We actually need to use the roll stiffness or roll rate, K_θ. The roll rate is the amount of torque required to produce one degree of roll in the sprung mass. The roll rate depends on the ride rate, K_R, which is the effective spring rate at the wheel. The ride rate depends on the tire sidewall spring rate, K_T, in series with wheel spring rate, K_W. The wheel rate depends on the spring constant, K, and on the spring location. Figure 5.32 shows a simplified sketch of a wheel, control arm, and spring.

Let's start with a known spring rate and work our way toward finding the roll stiffness. From figure 5.32 we see that the centerline of the wheel is a distance b from the pivot and the spring is a distance a from the pivot. We define wheel rate as the rate of a theoretical spring located where the wheel is attached to the control arm. It must produce a torque about the pivot point equivalent to the torque produced by the installed spring. Again from figure 5.32 we see that when the wheel swings through an arc length S_b, the point where the spring attaches to the control arm swings through an arc length S_a. The spring produces a force of magnitude KS_a and a torque $\tau_S = a\,(KS_a)$. Following a similar procedure for the wheel rate theoretical spring, K_W, and noting that the arc length is equal to the radius times the angle ($S = r\theta$), we have the following:

$$\tau_S = \tau_W$$

$$a(KS_a) = b(K_W S_b)$$

$$a(K(a\theta)) = b(K_W(b\theta))$$

$$K_W = K\left(\frac{a}{b}\right)^2.$$

The quantity (a/b) is called the motion ratio or the linkage ratio. The tire sidewall behaves like a spring with a spring constant K_T. We define the ride rate,

K_R, as the wheel spring rate in series with the tire spring rate. In section 5.8 we learned how to find the effective spring constant of springs in series:

$$K_R = \frac{K_T K_W}{K_T + K_W}.$$

Now that we have the ride and wheel rates, it is time to consider rolling of the chassis left or right as the car takes a corner. Rolling of the chassis is really a rotation of the sprung mass about the roll axis on the longitudinal centerline of the car. The wheel spring rate acts on the sprung mass at a distance from the centerline equivalent to half of the track, T, producing a torque on the sprung mass. The wheel on the opposite side of the car produces a similar torque on the other wheel. The sum of these two torques is equivalent to the roll rate, K_θ, times the roll angle, θ (note that roll rate is a torsional spring constant):

$$\tau_F = K_{F,\theta}\theta = 2(\text{Force})(\text{lever arm}) = 2\left((K_{F,R}S_F)\frac{T_F}{2}\right)$$

$$S_F = R_F\theta = \frac{T_F}{2}\theta$$

$$K_{F,\theta} = \frac{1}{2}K_{F,R}T_F^2.$$

The roll rate or roll stiffness is different at the front and rear of the car, so we add an F (front) or an R (rear) subscript to differentiate the two.

We are now in a position to discuss the torques that control the load transfer when the car accelerates around a corner. We will follow an abbreviated version of the approach of William and Douglas Milliken in their book *Race Car Vehicle Dynamics* as corrected in the companion text, *Race Car Vehicle Dynamics: Problems, Answers and Experiments*, by Milliken, Kasprzak, Metz, and Milliken. We begin by separating the sprung and unsprung mass and considering the torques generated in a turn, which generate load transfer. Let's assume that the car has left-right symmetry and that the roll axis remains fixed. Thomas Crahan in his SAE paper 94350, "Modeling Steady-State Suspension Kinematics and Vehicle Dynamics of Road Racing Cars Part 1: Theory and Methodology," points out that this is an oversimplification, and he corrects for both the vertical and lateral motion of the roll center. The motivated reader

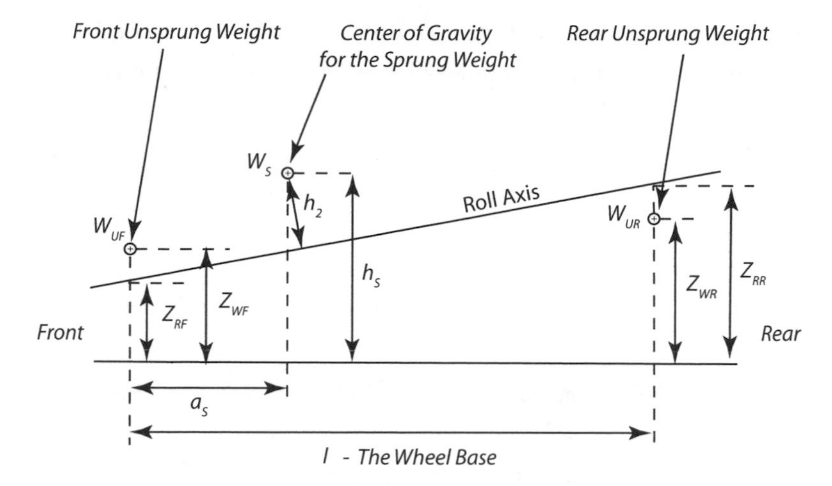

Figure 5.33 Centers of gravity and relevant heights and lengths used to evaluate front and rear load transfer due to cornering.

can find this paper included as a part of Carrol Smith's *Racing Chassis and Suspension Design*.

Figure 5.33 defines the relevant parameters that locate the centers of mass and the roll axis. The parameters are identified as follows:

- W_{UF} is weight at the center of gravity for the front unsprung weight.
- W_{UR} is for the rear.
- W_S is the weight at the center of mass for the sprung mass.

Keep in mind that none of these three are located at the center of mass for the entire car.

- Z_{RF} and Z_{RR} are the heights above the ground of the roll center at the front and rear axle, respectively.
- Z_{WF} and Z_{WR} are the heights above the ground of the front and rear unsprung centers of mass, respectively.
- h_2 is the perpendicular height of the sprung mass above the roll axis.
- h_S is the height of the sprung mass above ground.
- θ is the roll angle.
- A is the acceleration divided by the acceleration of gravity, g.

In the noninertial reference frame of the car, we treat the acceleration as an inertial force. There are four torques that make up the load transfer equation for the front wheels:

$$t_F \Delta W_F = \left[-Z_{WF} W_{UF} A\right]_1 + \left[\left(\frac{l - a_S}{l}\right) Z_{RF} W_S A\right]_2 + \left[\left(\frac{K_{F\theta}}{K_{F\theta} + K_{R\theta}}\right) h_2 W_S A\right]_3 + \left[\left(\frac{l - a_S}{l}\right) h_2 W_S \theta\right]_4 . \tag{5.21}$$

Term 1 is the torque generated by the inertial force of acceleration acting on the unsprung mass. Term 2 is due to the inertial force acting at the roll axis at a point directly beneath the center of gravity. The fraction in parentheses is the proportionality fraction for sprung weight that acts on the front wheels. Term 3 is the torque on the sprung weight about the roll axis due to the inertial force. The fraction in parentheses is the fraction of the roll stiffness that is due to the relative front roll stiffness. Term 4 is torque on the sprung weight about the roll axis due to gravity. Again, the term in parentheses proportions the torque according to the fraction of the sprung weight on the front wheels. This is the weakest torque of the three, and we will neglect its contribution. We find a similar torque relationship for the rear wheels:

$$t_R \Delta W_R = \left[-Z_{WR} W_{UR} A\right]_1 + \left[\left(\frac{a_S}{l}\right) Z_{RR} W_S A\right]_2 + \left[\left(\frac{K_{R\theta}}{K_{F\theta} + K_{R\theta}}\right) h_2 W_S A\right]_3 + \left[\left(\frac{-a_S}{l}\right) h_2 W_S \theta\right]_4 . \tag{5.22}$$

Again, we can neglect the gravitational torque contribution, term 4.

Once we have these two expressions, we can begin to consider the effect of changing the design parameters. For example, what happens if we raise the rear roll center, Z_{RR}? Term 1, the torque of the unsprung mass, is unaffected for the rear weight transfer in equation (5.21). Term 2, the torque due to the inertial force acting at the roll axis, increases linearly with the rear roll height. When the rear of the roll axis moves up, h_2, the sprung weight lever arm, decreases. The approximate relationship from comparing the arc lengths is

$$\Delta h_2 = -\frac{a_S}{l}\Delta Z_{RR}\,. \tag{5.23}$$

Replacing h_2 with $(h_2 + \Delta h_2)$ and Z_{RR} with $(Z_{RR} + \Delta Z_{RR})$, we can find the change in load transfer:

$$\text{Change in rear load transfer} = \Delta Z_{RR}\left(1 - \frac{K_{R\theta}}{K_{F\theta} + K_{R\theta}}\right)\frac{a_S}{l}\,W_S A\,. \tag{5.24}$$

The increase in load transfer at the rear decreases the net lateral force produced by the rear tires. In the front load transfer equation (5.21), only torque term 3 changes with Δh_2. Substituting equation (5.23) into the front load transfer equation (5.21), we have a decrease in load transfer at the front:

$$\text{Change in front load transfer} = -\Delta Z_{RR}\left(1 - \frac{K_{F\theta}}{K_{F\theta} + K_{R\theta}}\right)\frac{a_S}{l}\,W_S A\,. \tag{5.25}$$

The roll stiffness terms in parentheses in both equations (5.24) and (5.25) are equal. The front tires lose the load transfer gained by the rear tires. The decrease in the front load transfer increases front lateral force. Thus, we have found one way to correct for understeer: more front lateral traction and less rear lateral traction. In NASCAR terms, we have "loosened" the car. Lowering the rear roll center will "tighten" the car, or correct for oversteer.

5.11 CORRECTING HANDLING PROBLEMS

Crew chiefs in NASCAR can loosen or tighten handling conditions during a pit stop. A long ratchet wrench is inserted through a hole in the right side of the back window. The upper connection of the Panhard bar (fig. 5.17) can be raised. *This raises the rear roll center height. We saw in the last section that this corrects an understeer condition.* Lowering the Panhard bar upper mount can correct oversteer.

We now have enough information to apply a number of solutions to handling problems. This is not meant to be an all-inclusive list. Carroll Smith's book *Drive to Win* contains a more complete list, as does Milliken and Milliken's *Race Car Vehicle Dynamics*.

Designing, building, and correcting problems are at the heart of the Formula SAE program. Photograph by Clyde Caplan and Alex Teitelbaum.

5.12 UNDERSTEER CORRECTION

Lowering the front tire pressure by a half a pound or a pound will increase the size of the contact patch and reduce the spring rate of the tire sidewall. Increasing the contact patch will increase the lateral force at the front and *correct for understeer.* Lowering the tire sidewall spring rate, K_T, reduces the front roll stiffness and load transfer. This increases the front lateral force and corrects for understeer. The downside is that the reduced sidewall tension increases the steering angle at which the larger lateral force is achieved. The driver must turn the steering wheel farther to have the suspension stabilize. This can be very disconcerting to a driver. When I bought my first set of racing R-compound tires for autocross, I encountered a large slip angle tire for the first time. Some of the top drivers told me that the Kumho tires I had purchased operated better at lower pressures than the Hoosier autocross tires. Many drivers preferred the positive feel of the Hoosier tires, even though the Kumho tires produce a higher peak lateral grip. This changes from year to year as the manufacturers tweak the compound and construction of the tires. There is a limit to all good things. Further lowering of the pressure may produce distortion in the contact patch that reduces grip. There is always an optimum pressure. Keep in mind that there is a minimum operating pressure below which tire failure is a possibility. Race engineers look for uniform temperature across the tire as an indicator of optimum use of tire traction. Of course, higher inside temperatures are common when employing extensive negative camber. Extremely low pressure causes heating of the tire to the point where the tire can delaminate and fail catastrophically. Consult the manufacturer's guidance regarding safe operating pressures.

Understeer can be corrected by decreasing the front roll stiffness, $K_{F\theta}$. Decreasing the front roll stiffness decreases the front load transfer, which increases the overall front lateral force (grip). There are several ways to accomplish this. We can reduce the front spring rate by *replacing springs*. Spring rates are typically adjusted in something like 10% increments. In NASCAR the teams frequently use spring rubbers. These are small blocks of stiff rubber that are placed between loops of the coil spring, effectively removing that loop from compressing. The remaining loops must compress a little bit more for any given compression distance. This raises the spring rate. Understeer can be improved by *removing front spring rubbers*. We can *reduce the front antiroll bar stiffness* by replacing or adjusting the bars. Adjustable antiroll bars can have a series of holes drilled in the two arms (fig. 5.18). Selecting a set of holes farther from the base of the U increases the lever arm for the suspension and decreases the effective stiffness of the bar.

If front roll stiffness is too low, the suspension can literally slump over on the bump stops on the outside when making a turn. With no more suspension travel, behavior becomes erratic. In this circumstance, we must increase the front roll resistance. We must choose to increase either the spring rate or the antiroll bar rate. Increasing the antiroll bar rate will compress the suspension for the inside tire. Increasing the spring rate will not. On the other hand, if the track is bumpy, especially in the corners, increased spring rates will decrease traction. A soft spring rate is needed for the tires to comply with the track surface. Antiroll bars will increase the front roll stiffness without adversely affecting compliance with bumps.

Increasing the front low-speed bump force in the shock absorbers can reduce corner entry understeer. Corner entry involves two actions that transfer load onto the outside front tire. The first is the turn itself, and the second is braking. If the transition happens too quickly, we can induce suspension oscillations that reduce traction. Increasing the front shock low-speed bump rate will help to control the suspension transition rate and provide improved stability during corner turn-in. It is important to keep in mind that once the suspension is "set," the shock velocity goes to zero and the shock absorbers no longer play a role. Mid-corner handling problems cannot be addressed with shock absorbers.

Improper front suspension geometry can create understeer. For example, if the front wheel camber goes positive under bump, the wheel will lose traction. Bump steer, where the steering direction changes as the suspension compresses, can create understeer. Correcting suspension geometry can involve a significant amount of work.

Wheel alignment can be responsible for understeer. *Increasing front negative camber can improve understeer.* Negative camber provides camber thrust, which helps corner turn-in. It can also help keep the tire upright under load, which increases the contact patch area and lateral force. Excessive toe-in or toe-out can reduce the lateral force of both tires and can also create premature wear. A small amount of toe-in creates high-speed stability. A small amount of toe-out can increase the rate of turn-in for autocross, but frankly, it is dangerous if driven on public roads.

NASCAR uses another suspension adjustment called wedge. The team changes the upper rear spring perch height by sticking a ratchet through another hole in the rear window. This adjustment changes the cross weight of the car. For example, if we push down on the right rear spring perch (which raises the right rear ride height), we increase the load on the right rear tire. This is equivalent to sticking a matchbook under a table leg. The right rear and left front tires gain vertical load. The left rear and right front lose vertical load. *Increasing right rear wedge improves lateral force at the front and decreases lateral force at the rear and corrects for understeer. Beware: this only works because* NASCAR *only turns left!*

Understeer can be improved by replacing the tires. Race compound tires have a limited lifetime. They require an initial low-stress thermal cycling to relax some of the high-stress bonds in the rubber and allow the shedding of mold release compounds and other volatile residual chemicals. Finally, the tires are allowed to cool and anneal for 24 to 48 hours and form new higher strength bonds (see Paul Haney's *The Racing & High-Performance Tire: Using the Tire to Tune for Grip and Balance*). Some manufacturers and tire suppliers will, for a price, heat cycle the tires on test rigs before delivery. The first few laps after this process will typically be as good as the tire gets. With each heat cycle of the tire, the tread compound gets harder and less adhesive and compliant. A well-designed autocross car will use all of the available heat cycles long before the tire wears out. An amateur racer will have to shell out between $600 and $1300 for a set of R-compound tires. It is hard to decide that a set of tires are "gone" at these prices. I can't imagine how drifters feel when they burn through two $300 rear tires in 15 or 20 minutes. Would-be amateur drifters will drive on any piece of rubber that fits on the rim. The harder the tire compound, the better. But once you commit to serious drifting, the quality of the tire matters. I only bring this up to make myself feel better about my own racing expenses.

So far, we have concentrated on improving the traction situation at the front wheels. This is the preferred method of curing understeer. *Always try to improve traction on the end of the car that is weakest.* This isn't always possible. Sometimes the rules of the racing class prevent the action you would prefer to take. Sometimes the parts are not available. Sometimes, to restore balance, we must work on the opposite end of the car. For example, *corner entry understeer can be improved by increasing the rear roll stiffness.* This will increase the rear load transfer and increase the rear slip angle. This improves the rotation rate of the car.

5.13 OVERSTEER CORRECTION

After going through understeer correction, the list here will seem familiar. *Oversteer on corner entry can be corrected by reducing the car's rear roll stiffness.* This reduces rear load transfer and improves the rear lateral force. We can accomplish this by reducing the rear spring rate or antiroll bar rate. Less rear weight transfer means more rear lateral force.

Oversteer on exit can be corrected by reducing the rear roll stiffness. This reduces rear weight transfer and improves the overall rear traction. Rear roll stiffness is reduced by replacing the spring and adjusting the antiroll bar.

Oversteer on exit can be improved by increasing the rear shock low-speed bump. Oversteer on exit is subject to two actions. First are the turning forces, which transfer load to the outside, and the second is due to stepping on the gas. This causes a rearward weight transfer. These forces need to be controlled during the transition period. Increasing the rear shock low-speed bump stiffness can stabilize the rear suspension on corner exit and improve rear traction. It's always possible to get too much of a good thing. Too much rear low-speed bump transfers too much load to the outside tire and leads to oversteer and wheel spin when you step on the gas as you exit the corner.

On a rear-wheel-drive car, a worn-out limited slip differential can cause oversteer on corner exit and an inability to "put down the power." The limited slip differential's job is to apply torque to both rear wheels. As it wears out, torque is transferred in an erratic fashion between the wheels with a preference for the inside wheel. Typically, this leads to inside wheel spin and oversteer.

Reduced rear tire pressure can improve rear traction and combat corner exit oversteer and wheel spin. Less tire pressure means a larger contact patch, a softer sidewall spring rate, and more traction.

A stiffer front antiroll bar can help with corner-exit wheel spin. This is a diagonal weight transfer effect. The stiffer bar increases the front load transfer to the outside front wheel. This increases the load on the inside rear wheel and reduces the wheel spin of this wheel.

Finally for us, *worn-out R-compound tires can lead to wheel spin and oversteer on corner exit under throttle.* Tracking durometer hardness measurements of your tires over their life can help you to spot this transition to a tire that has run out of heat cycles.

Once again I want to point out that this list is incomplete. We have yet to consider the effects of aerodynamics or suspension travel. Insufficient bump or droop (compression or rebound) suspension travel can create all kinds of handling problems. Load transfer occurs very quickly as the suspension runs out of travel. Next time you see a race broadcast, watch the adjustments made during pit stops. What adjustments did they make? What reasons did the announcer provide? Can you take your newfound knowledge and follow the crew chiefs' decision process?

5.14 SUMMARY

We started the chapter with goals of developing an understanding of steering, suspensions, springs, and dampers. We also wanted to develop approaches to correcting handling problems. Here are the highlights of what we learned:

- Ackermann steering geometry is necessary in a turn to compensate for the difference in the radius of curvature for inside and outside wheels.
- Slip angle is the difference between the direction a tire is pointed and the direction of the instantaneous velocity vector.
- Oversteer is characterized by a larger slip angle at the rear than at the front. Understeer is the opposite.
- Turning the steering wheel creates an induced tire drag that slows the car. Less turning of the wheel means faster track times.
- Negative camber produces camber thrust, which aides in corner turn-in. It also improves tire contact with the road under heavy turning and improves tire wear under these conditions.
- Caster creates a self-aligning torque that tends to keep the car heading straight. As the steering angle increases, the self-aligning torque begins to decrease. The driver feels this in the steering wheel and can use it as an indicator of the onset of understeer.
- Toe-in can help high-speed stability. Toe-out can aid rapid corner turn-in at low speed and instability at high speed. Too much of either is bad.
- Wheel changes that increase the scrub radius make the steering feel heavy.
- A double A-arm suspension with a shorter upper control arm adds negative camber under loading.
- A MacPherson strut front suspension is simple, adjustable, and works well with front-wheel drive.
- We learned a little about adjustability in the NASCAR-type solid rear axle.
- Lowering the roll center keeps the chassis level. If the roll center falls

below the road surface, a jacking force is created that lifts the chassis under turning.

- Sway bars transfer vertical load to the outside tire and keep the chassis level.
- We modeled cars as simple harmonic oscillators. Dampers must control oscillations at the resonant frequencies. Typically, there are two resonances that are associated with the sprung and unsprung weights.
- Dampers that are adjustable in bump and rebound are used to correct handling problems.
- Cornering forces generate four torques that determine the amount of vertical load transfer between the wheels. We adjusted parameters in the torque equation to correct handling problems.
- Finally, we considered a variety of handling problems and the tools to correct them.

Chapter 6

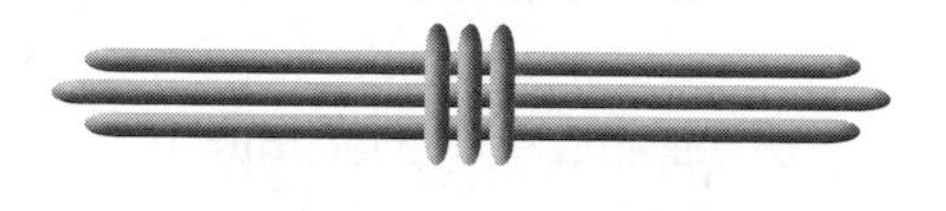

Green Racing

To this point, we have considered many of the factors that make race cars work and drivers win. It is time to shift our focus to the future and look at potential new technology. We call this future "green racing," and at its heart is the study of energy.

The most fundamental need in the universe is energy. Without energy in the proper form, nothing happens. The sun doesn't shine, chemical processes do not occur, and life does not exist. Most processes that use energy to perform a task end up wasting some of it. They leave behind unreacted fuel, end-product chemicals, and excess thermal energy. It is the nature of thermodynamic processes. To exist in equilibrium, any system must find a source of useable energy and a way to deal with the waste. This is true for a fish tank, a submarine, a space station, or a planet. Ignore these needs and your system is doomed. Cars in our environment are no exception. Primarily they take in energy stored in the form of petroleum products and give off smaller hydrocarbon chains, carbon monoxide and carbon dioxide, nitrous oxides, and waste heat. The waste heat is typically a staggering 70% of the energy consumed.

Petroleum, a near-perfect gift from the planet, has been easy to find, high in energy density, and nearly limitless in its uses. After a century and a half of exploiting petroleum, it is clear that it is not limitless in quantity. As our planet's industrial population grows, the growth rate in oil extraction from the ground has begun to decrease. Best estimates indicate that we are approaching the peak in extraction rate. Once we achieve the peak rate, the cost of oil will begin to skyrocket as the industrial economies compete for oil. At the same time, the effects of waste products are beginning to overwhelm the processes that keep them in check. Pollution control systems, when properly employed, can deal with most of the carbon monoxide, nitrous oxides, and unburned hydrocarbons. Carbon dioxide is another story. We count on plant life to absorb and sequester the carbon dioxide, converting it back to hydrocarbons. As humans deforest a bigger and bigger fraction of the planet's surface, the ability to remove CO_2 is degraded. The big picture is not rocket science. The excess carbon dioxide enhances the atmosphere's greenhouse effect. Sunlight streams through our transparent atmosphere, just as it does with the glass of a greenhouse. That light is composed of a wide range of wavelengths. Absorbed by the land and sea, the light energy is then reemitted, primarily in the wavelengths of infrared light. Greenhouse glass and greenhouse gases are opaque to infrared light and trap the energy. Thus, as the greenhouse gas content in the atmosphere rises, it throws the system out of balance and the planet begins to warm.

In the face of rising energy demand, fuel prices, pollution, and global warming, will racing survive? I believe the answer is yes. Humans have a fundamental urge to compete. We race on foot, on horseback, on camels, on bicycles, and, if the boss isn't around, we'll even race on office chairs. Richard Petty said it best: "There is no doubt about precisely when folks began racing each other in automobiles. It was the day they built the second automobile." Automotive racing is entertainment, and as long as there are a few discretionary dollars in the budget, it will continue. Given our nature and our problem, it seems almost obvious that it is time for a Green Racing initiative. It is once again time to use racing to improve the breed.

Racers have always been resourceful. Some of the finest automotive engineering has been developed for racing. That engineering often finds its way into street cars. Ferrari has developed dozens of technologies from carbon ceramic brakes to manual electrohydraulic clutches that make their street cars cutting edge. McLaren Racing Technologies is an entire division of one of the most successful Formula 1 teams, whose mission is to assemble racing technologies and adapt them to other applications. Some racing series have already begun to tap into this creativity to make racing more ecologically friendly. Formula 1 included regenerative braking as an allowed technology in 2009. The Indy Car series has used alternative fuel for a number of years. Audi, a German manufacturer, won the 24 Hours of Le Mans in 2006 in the diesel-powered R10 TDI. It was the first time that a diesel-powered car had ever won this race. Audi previously used their R8 endurance racer as a test bed for their direct fuel injection technology. Direct injection is now widely used and has produced improved fuel efficiency, more power, and reduced pollution. Peugeot has announced their intention to race the 908HY hybrid car at the 2010 24 Hours of Le Mans. The race to be green is already underway.

6.1 WHAT IS GREEN RACING?

The most serious attempt to make a racing series environmentally friendly is the effort by the American Le Mans Series or ALMS. In partnership with the Environmental Protection Agency, the Department of Energy, and the Society of Automotive Engineers, the ALMS has begun to implement a "green racing initiative" called the Green Challenge. The first test of the initiative occurred in October 2008 at the Petit Le Mans at Road Atlanta in Braselton, Georgia. Corvette Racing, the factory team of General Motors, won the inaugural Green Challenge in the GT class, and Penske Racing's factory-sponsored Porsche RS Spyder won it in the Prototype class. In 2009, ALMS awarded championship points based on the results of the Green Challenge. This kind of initiative is only possible in a racing series that allows and, in fact, encourages innovation. A spec car series, such as NASCAR, where all platforms are essentially identical, will have a tough time contributing to new technology. On the other hand, if the rules are sufficiently flexible, racers can do a lot to improve overall efficiency. When the ALMS chose to establish the Green Challenge, they first had to decide what it meant to be "green." The answer is complex, and ignoring that complexity can lead to the wrong conclusions. The factors are represented in figure 6.1.

Several factors contribute to being green, efficient, and planet friendly. The

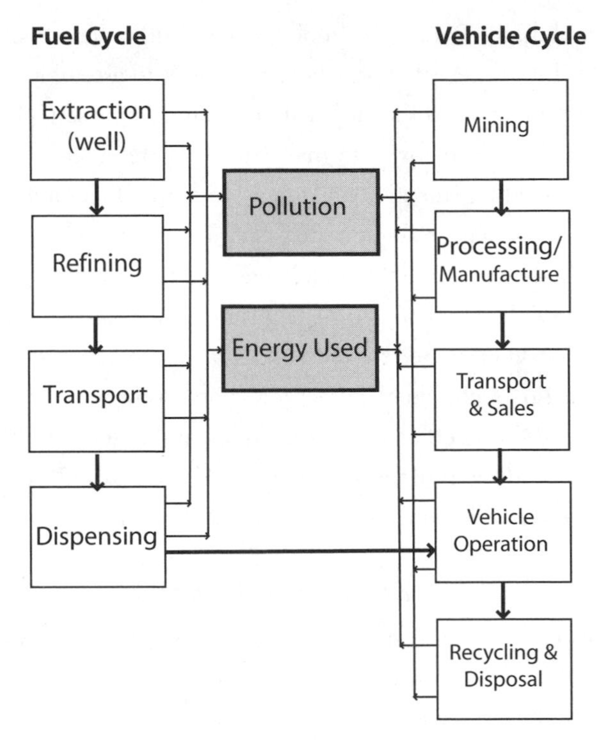

Figure 6.1 GREET model total energy cycle for transportation. A complete assessment must account for pollution, including greenhouse gases, and energy usage at all stages of fuel and vehicle production.

first element is controlling energy consumption. We can divide the energy consumption into two parts. The obvious part is the energy used directly by the car. This is a part of the puzzle where racing teams can make a significant contribution. The second piece is the energy used in producing the fuel. For petroleum, this is the energy used in the refining process. For example, if a car were powered by corn-based ethanol, we would include the energy required to grow and process the corn.

The second element is controlling the emission of greenhouse gases. This includes emissions from the vehicle and emissions from the fuel and vehicle production.

The third factor is the amount of petroleum fuel used, including that used by the race car and that used in fuel production. The idea is to reward the use of alternative fuels and reduce dependence on petroleum. A good example is to compare the use of petroleum with corn-based ethanol. Burning petroleum releases carbon as CO_2. Locked up for millions of years, this carbon is now a greenhouse gas. On the other hand, corn captures CO_2 from the atmosphere and energy from the sun. Distill the corn into ethanol. When the ethanol is burned, CO_2 is released back to the atmosphere. The net change in this atmospheric greenhouse gas is zero. Burning a food source may not be the best idea, but the principle can be extended to other plant sources.

Argonne National Laboratory's Transportation Technology R&D Center developed the GREET model that is used to assign scores in the Green Challenge. GREET is the Greenhouse gases, Regulated Emissions, and Energy use in Transportation model. The GREET model is designed to give researchers a basis for evaluating vehicle and fuel combinations for the full fuel cycle and vehicle life cycle. The GREET simulation program may be downloaded from the transportation section of the Argonne National Laboratory Web site at www.transportation.anl.gov. Notice that the GREET model worries about disposal of the vehicle at end of its life. How do you dispose of your car if it is a toxic nightmare?

The International Motor Sports Association (IMSA) governs the ALMS. At the moment, under IMSA rules, teams are limited to one of three fuels. IMSA 100E10 is a gasoline that is 91% petroleum. IMSA E85R is an ethanol fuel that is 34% petroleum. Shell GTL is a diesel fuel that is 95% petroleum based. The ALMS organization then assigns engine displacements and minimum weights to level the playing field on the racing side of the competition. Gas tanks are sized to place equal amounts of energy on board vehicles in the same class.

In summary, being truly green is complicated. To make a comparison, we must consider the entire life cycle of the fuel and the vehicle. We must focus our research and engineering on the weak point of each. Once armed with the full picture, we can we make a commitment to a particular technology.

Let's consider some of the kinds of technologies that racers could use and develop if the rules were unlimited.

6.2 REGENERATIVE BRAKING

Conventional disk brakes, shown in figure 6.2, use friction between nonrotating pads and a rotating disk to slow the car. The kinetic energy of the car ends up as thermal energy in the pads and the rotors. This is bad for two reasons.

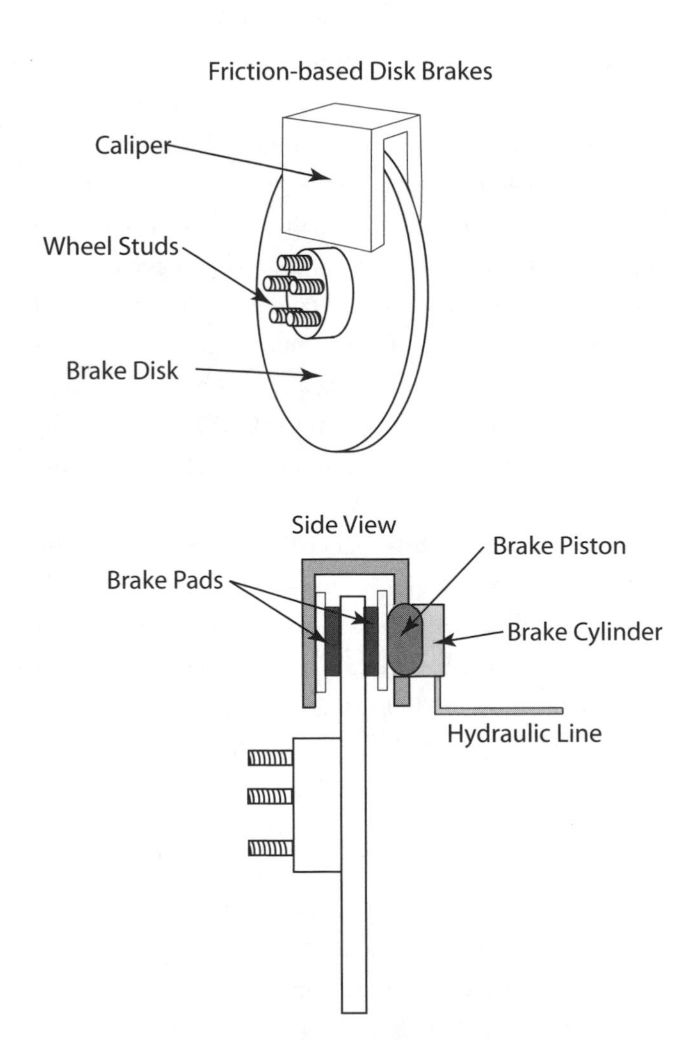

Figure 6.2 Conventional disk brakes. Hydraulic pressure pushes the piston against the brake pads. Friction between the brake pads and the disk converts kinetic energy to thermal energy.

First, we must shed the thermal energy to the environment. Failure to do so will result in overheating and brake failure. Second, the energy is lost and unusable. Regenerative braking minimizes this loss by avoiding friction-based braking.

Hybrid passenger cars already employ regenerative braking to some degree. In this technique we convert the vehicle's kinetic energy, the energy associated with motion, to some form of stored energy. The most insightful way to approach this problem is to consider the conservation of mechanical energy equation. Conservation of energy is one of the most fundamental principles in all of physics. It says that energy can neither be created nor destroyed. It simply changes forms. Just as an accountant must keep track of every penny in a ledger, a physicist must account for every joule of energy in a system. We will consider our car to be our system. Let's start with the simple case in which our system is isolated and temperatures are constant. We then look to conserve mechanical energy. Mechanical energy is made up of two parts, kinetic energy, *KE*, and potential or stored energy, *PE*. *KE*, as we said above, is the energy of motion. If the center of mass moves, then $KE = \frac{1}{2} mV^2$, where m is the mass and V is the speed of the center of mass. We call this translational *KE*. If the motion is rotation, then $KE = \frac{1}{2} I\omega^2$, where I is the moment of inertia and ω is the angular velocity (see chap. 1). We call this rotational *KE*. In physics class, the first stored energy case studied is gravitational potential energy, where $PE = mgh$ and m is mass, h is the height above some reference, and g is the acceleration of gravity. A simplistic classroom model of regenerative braking would be to use a hill. If a moving car rolls up a hill, gaining *PE*, it slows down and loses *KE*. As the car crests the peak and rolls down the other side, it regains its kinetic energy as it heads downhill and loses potential energy. If the i subscript represents the initial conditions and f is final, we can write conservation of energy as

$$KE_i + PE_i = KE_f + PE_f$$

or

$$\frac{1}{2}mV_i^2 + mgh_i = \frac{1}{2}mV_f^2 + mgh_f. \qquad (6.1)$$

If $h_f = h_i$, all of the kinetic energy is recovered and we don't need to expend fuel to regain our speed. For simplicity, we have neglected the energy lost from rolling resistance in the tires and aerodynamic drag.

While this is technically correct, it is not terribly practical. Conventional brakes will be far more practical than looking for a hill to stop the car. In conventional brakes, we press ceramic or semimetallic brake pads against metallic rotating disks. Kinetic friction between the pad and disk converts kinetic energy to thermal energy, ΔE_{th}. In other words, the rotor and pads get hot. They convert a great deal of kinetic energy to thermal energy and in turn depend on cooling air and infrared radiation to carry the thermal energy away. We call this non-regenerative braking because the energy is lost from our system. Our conservation of energy equation must also account for the thermal energy:

$$\frac{1}{2}mV_i^2 + mgh_i = \frac{1}{2}mV_f^2 + mgh_f + \Delta E_{th}. \quad (6.2)$$

Assuming level ground ($h_i = h_f$), the final speed is always less than initial speed and fuel must be expended to restore the kinetic energy to its initial value.

6.3 MECHANICAL ENERGY STORAGE: FLYWHEELS

Some early attempts at regenerative braking involved the use of a large heavy rotating disk, called a flywheel, to store the energy in the form of rotational motion. We should add a rotational kinetic energy term for this case:

$$\frac{1}{2}mV_i^2 + mgh_i + \frac{1}{2}I_{\text{flywheel}}\,\omega_i^2 = \frac{1}{2}mV_f^2 + mgh_f + \Delta E_{th} + \frac{1}{2}I_{\text{flywheel}}\,\omega_f^2.$$

For level ground, the *mgh* terms drop out. If we avoid conventional brakes and other resistive losses, the ΔE_{th} can be neglected and we are left to trade energy between translation of the car and rotation of the flywheel.

This idea faces a number of technical challenges. The added weight of the flywheel results in efficiency losses in the system. The large rotating mass also creates a large angular momentum ($L = I\omega$) and large gyroscopic torques, which interfere with the handling of the car. This is the same phenomenon experienced in front-wheel-drive cars and is known as "torque steering." This can be addressed by using two counter-rotating flywheels, effectively cancelling the net angular momentum. We also must deal with the fact that rapidly rotating masses produce large internal stresses in the flywheel's metal that can lead to catastrophic failure. All of these problems, combined with the complexity of adding and removing energy from the flywheel, have prevented us from employing it in most practical applications. As material technologies have improved, some are rethinking the flywheel approach. Smaller masses rotating at very high rates can store substantial energy (proportional to ω^2) while minimizing the angular momentum (proportional to ω). Several Formula 1 teams are trying such flywheels for use in kinetic energy recovery systems. Let's consider a Formula 1 example where a 605 kg (1330 lb) car slows from 150 mph to 75 mph. How much translational *KE* in joules must be absorbed? A value of 150 mph is approximately 67.1 m/s, and 75 mph is 33.5 m/s. Thus,

$$\Delta KE = \frac{1}{2}mV_f^2 - \frac{1}{2}mV_i^2 = \frac{1}{2}(605)(33.5)^2 - \frac{1}{2}(605)(67.1)^2 = -1.02 \times 10^6 \text{ J}.$$

Roughly a million joules of energy must be absorbed. Let's assume that the flywheel is a 10 kg disk with a radius of 10 cm and that the disk is not rotating initially ($\omega_i = 0$) and is spun up as the car slows. The moment of inertia, I, of a disk is equal to ½ mR^2, where m is the mass of a disk and R is the radius:

$$\frac{1}{2}I_{\text{flywheel}}\,\omega^2 = \frac{1}{2}\left[\frac{1}{2}10(0.1)^2\right]\omega^2 = 1.02 \times 10^6 \text{ J}$$
$$\omega = 4520 \text{ rad/s} = 43{,}100 \text{ rpm}.$$

The details of the Formula 1 flywheels are proprietary, but the suppliers hint at angular speeds up to 100,000 rpm. To minimize losses, the flywheel typically spins in a vacuum.

6.4 ELECTRICAL ENERGY STORAGE: BATTERIES

We have considered storing energy using gravity and mechanical means, but what about electrical storage? A small motor/generator connected to the car's driveline can convert mechanical energy into electrical energy. Structurally there is little difference between an electrical motor and an electrical generator. With a little careful design, electrical engineers can provide a generator that converts mechanical energy into electrical energy under "braking," as well as becoming a motor that converts electrical energy into mechanical when it is time to speed up. This energy can be stored in a number of ways. One way is to use a battery. We can divide batteries into two broad classes, those that are rechargeable and those that are not. Both classes provide electrical energy that has been stored in chemical bonds. Some chemical reactions are reversible and some are not. If the battery relies on a reversible chemical reaction, the battery is rechargeable. Perhaps the most common rechargeable type is the lead-acid battery found in almost every car and truck in the country. It is simple, reliable, and has been used in cars for almost a century. The principles behind its function are similar to all batteries. We will study its function as a model for understanding other batteries and even fuel cells.

A battery consists of two electrodes and an electrolyte (fig. 6.3). A lead (Pb) electrode and a lead oxide (PbO_2) electrode reside in a water–sulfuric acid (H_2SO_4) bath. The sulfuric acid ionizes in the water and splits into protons (H^+) and HSO_4^- ions, forming a conducting electrolyte solution. The Pb plate combines with the HSO_4^- and forms lead sulfate ($PbSO_4$), protons (H^+), and two electrons ($2e^-$). The protons are transported through the electrolyte to the PbO_2 side of the cell. The electrons collect on the lead plate, establishing an electric potential (voltage) for the battery. External to the battery we connect the lead electrode to the wires of an electrical circuit and back to the lead oxide plate. The electrons flow through the wires, transferring energy to the circuit. The electrons return to the lead oxide plate. There they combine with the protons and HSO_4^- from the electrolyte and form lead sulfate and water. The overall reaction is as follows:

$$Pb + PbO_2 + 2\ HSO_4^- + 2\ H^+ \rightarrow 2\ PbSO_4 + 2H_2O \text{ Discharging.}$$

This reaction produces an electric potential of about 2 volts between the two electrodes. A 12 volt car battery requires six of these cells connected in series. As the battery cell provides electric energy, the chemicals are consumed, leaving lead sulfate. If we have an electric potential source that is slightly greater than 12 volts, we can reverse the chemical reaction and recharge the battery.

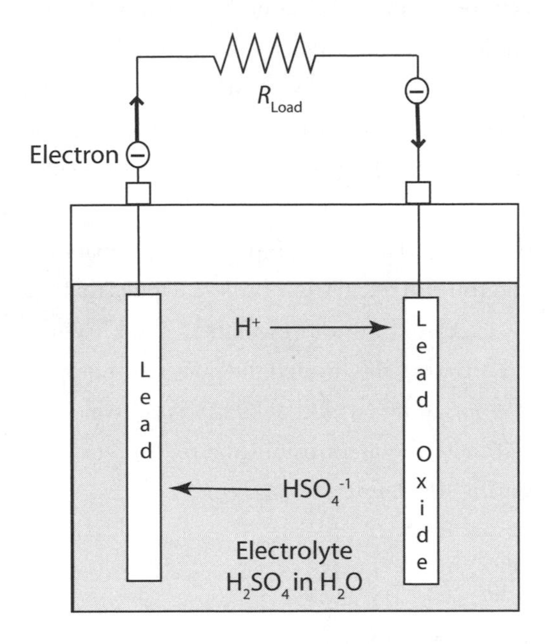

Figure 6.3 A single-cell lead-acid battery and external circuit.

The amount of stored energy is scalable up to vast sums. In nuclear submarines, a room-sized lead-acid battery bank serves as the ship's backup power supply. While weight is not an issue for a submarine, a large-scale lead-acid battery is unsuitable for racing or electric propulsion in cars.

Another common rechargeable battery is the NiCd or nickel-cadmium battery. It has nickel-hydroxide and cadmium electrodes with a potassium-hydroxide electrolyte. It suffers from a loss of energy capacity that depends on the history of its charge and discharge cycles. The cadmium is also toxic, which creates a disposal problem for large-scale applications.

We know that we want a low-weight battery. What other characteristics should our battery have? If the battery occupies too large of a volume, the outer shell of our race car will have to be larger. A larger shell means more aerodynamic drag. High energy and small volume mean that a high energy density is critical. Formula 1 has limited the stored energy to 400 kJ. A lead-acid battery has an energy density of approximately 150 kJ/kg. This would mean a 2.67 kg battery in a sport that worries about grams, let alone kilograms. The two most popular rechargeable batteries for large-scale use are the nickel-metal hydride (NiMH) battery and the lithium-ion battery. A NiMH battery has an energy density of approximately 340 kJ/kg, and a lithium-ion battery has an even greater energy density of around 460 kJ/kg. This would be a net savings of almost 2 kg.

In a kinetic energy recovery system, the battery must also be capable of charging and discharging quickly. You will recall from earlier chapters that the rate of energy flow is power. It is expressed in units of watts in the SI system and horsepower in British units. Formula 1 has limited the power output from the stored energy system to 60 kW for 2009. A value of 60 kW is equivalent to approximately 82 hp. We can estimate the amount of time that the car can supply boosted power by considering the definition of power:

$$\text{Power} \equiv \frac{\text{Energy}}{\text{time}}$$

$$\text{time} = \frac{\text{Energy}}{\text{Power}} = \frac{400{,}000 \text{ J}}{60{,}000 \text{ W}} = 6.66 \text{ s.}$$

So, for 6.66 s out of every lap an 800 hp car becomes an 882 hp car, a fantastic advantage when drag racing down the main straight attempting a pass. This 10% boost to engine power is clearly useful. But how helpful is the regenerative braking? We showed in our flywheel example that an F1 car must shed around 1 MJ in slowing from 150 to 75 mph, which is more than twice the capacity of their storage system. We must also consider the rate at which the regenerative system can absorb power. A Formula 1 car brakes at more than 4 g's. How much power is absorbed when braking at this rate?

If the car moves in the *x*-direction, we can write

$$\text{Power} = (\text{Force})_x \, (\text{velocity})_x \, .$$

Using Newton's second law ($F = ma$),

$$\text{Power} = (\text{mass})(\text{acceleration})_x \, (\text{velocity})_x \, .$$

Braking an F1 car at 4 g's from 150 mph (67.1 m/s), we have

$$\text{Power} = (605 \text{ kg})(4)(9.8 \text{ m/s}^2)(67.1 \text{ m/s}) = 1.59 \text{ MW}.$$

A value of 1.59 MW is an incredible 2170 hp! At top speed and with optimum braking, this value can climb as high as 2600 hp. Regenerative braking, at 82 hp, is not going to significantly help with slowing the car. Massive carbon ceramic friction brakes will do the bulk of the work. This story changes when you consider a passenger car. A 3000 lb car traveling 60 mph braking at 0.5 g needs 179 kW of braking power. A 60 kW system could supply a third of the required braking. The total kinetic energy in this car is about 491 kJ. While this is more than the capacity of an F1 car, it is roughly 10% of the capacity of the Toyota Prius battery. Regenerative braking has the potential to be very effective in a passenger car. In fact, it is common for owners of the Prius to report having friction brakes that last more than 100,000 miles. This is 2 to 3 times the typical lifetime of friction brakes and is evidence that regenerative braking makes a big difference.

6.5 ELECTRICAL ENERGY STORAGE: CAPACITORS

Capacitor storage of energy shares many of the features of battery storage. They have one distinct advantage in that most standard capacitors can charge and discharge very quickly. That is because the energy is stored differently.

We saw that batteries depend on chemical reactions to release the energy or recharge the battery. The energy is stored in the chemical bonds. Capacitors store energy by creating a charge distribution, a much faster process. When an external voltage source is applied across a capacitor, it moves charge from one conducting plate of the capacitor to the other. Figure 6.4 is a qualitative drawing of a capacitor being charged by a battery.

Both conducting plates in the capacitor start in a neutral state with zero net charge. It is neutral because each plate contains an equal number of electrons and protons. Frozen in position, the nucleus of each atom in the plate contains the positive protons. A cloud of negative electrons surrounds the nucleus. The cloud contains the same number of electrons as the nucleus contains protons. In metallic conductors, approximately one electron per atom is free to move. When we apply an external voltage between the two conducting plates, the resulting electric field moves electrons through an external circuit from one

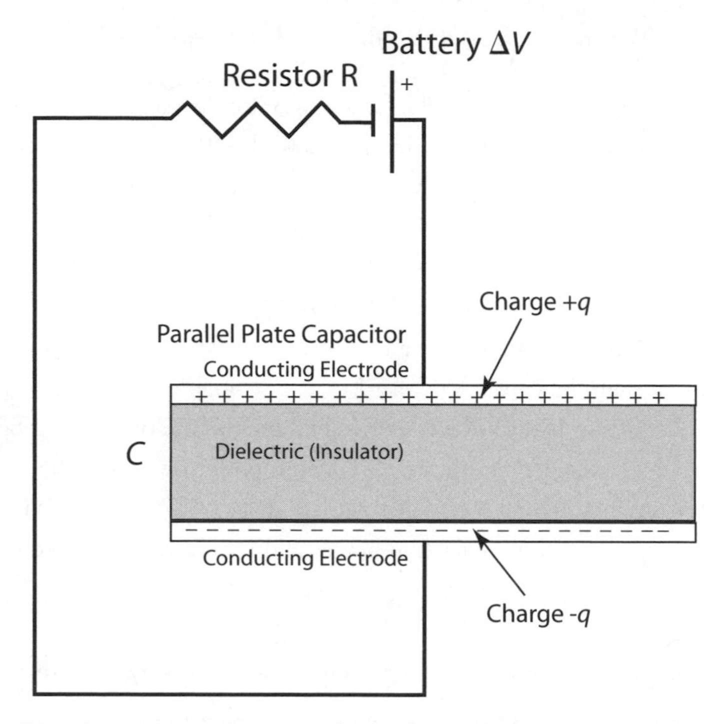

Figure 6.4 A conventional capacitor being charged by a battery with a resistor in series. When the battery and resistor are removed, potential energy remains stored in the electric field of the capacitor.

plate to the other and creates the charge distribution. We define capacitance as the ratio of the absolute value of the charge moved, q, to equilibrium voltage, ΔV, applied across the capacitor:

$$C \equiv \frac{q}{\Delta V}.$$

The resulting positive and negative charge distribution produces its own electric field. Energy is stored in this electric field. Adding an insulator between the conducting plates raises the amount of charge that can be stored, as well as the capacitance. The insulator, also called a dielectric, contains either permanent or induced electric dipoles. A dipole is a localized separation between positive and negative charge associated with the atoms or molecules that make up the dielectric. The dipoles are fixed in place, but they may rotate and change orientation with respect to the applied electric field. The reorientation of the dipoles due to the electric field of the conducting plates allows more charge to be stored on the capacitors for our given applied voltage. This in turn raises the capacitance, C. The potential energy stored in a capacitor is

$$PE = \frac{1}{2}C(\Delta V)^2 .$$

The reorientation of the dipoles and the flow of charge into and out of the capacitor is a very rapid process. Conventional capacitors are great at supplying pulses of power. Their downfall is a low volume energy density at around 3 J/cm^3, or roughly 1% of a lead-acid battery. The picture improves slightly when we consider the mass energy density. At 2 kJ/kg, the capacitor mass energy density is roughly 2% of that found in the lead-acid battery. Since we know that the energy stored is proportional to applied voltage squared, you might be tempted to raise the applied voltage. The problem is that the higher voltage increases the electric field within the capacitor. Every insulator has a limit to electric field that it can withstand. This limit is called the dielectric strength. Ultimately, the dielectric strength determines the practical energy density. A great deal of effort and money is being invested to address this limit.

Supercapacitors have energy densities that are an order of magnitude better

TABLE 6.1
Approximate Energy Density for Various Energy Storage Media

Storage Media	Energy/Volume	Energy/kilogram	Comments
Capacitors			Millions of cycles
Standard capacitors	3 J/cc	2 kJ/kg	Fast response
Supercapacitors	50 J/cc	3 kJ/kg	Slow response
Batteries			Thousands of cycles
Lead acid	0.300 kJ/cc	100 kJ/kg	
NiMH	1.10 kJ/cc	340 kJ/kg	Slow response
Li ion	0.830 kJ/cc	460 kJ/kg	Slow response
Hydrogen			
H_2 gas at STP	11 kJ/cc	143 MJ/kg	
Petroleum			
Gasoline	34 kJ/cc	46 MJ/kg	

than conventional capacitors. In a sense, a supercapacitor is a combination of a conventional capacitor and a battery. In these devices, one of the metal electrodes is replaced with a chemically active material. Because of the nature of the reaction and the proximity to the electrode, the reactions are still fairly fast and the energy density is better than conventional capacitors, but significantly less than that of batteries. Energy densities for various systems are summarized in table 6.1.

6.6 WHAT TYPE OF HYBRID IS IT?

An electric motor helping an internal combustion engine to propel a car is an example of a hybrid power plant. To simplify our language here, we will use the term "engine" when we mean the internal combustion engine and the term "motor" for the electric motor. (We can replace the engine with a fuel cell in any of these hybrid definitions. We will come back to the fuel cell shortly.) If the electric motor and the engine both provide propulsive force at the same time, we refer to the vehicle as a *parallel configuration hybrid* (fig. 6.5a). A Formula 1 car using regenerative braking is an example of the parallel configuration. When the only source of electrical energy for the motor is a storage device like a battery, we refer to the car as an *electric vehicle* (fig. 6.5b). Perhaps the best-known example is the Tesla Roadster from Tesla Motors. If the power for the electric motor comes from an engine driving an electrical generator, the vehicle is called a *series configuration hybrid* (fig. 6.5c). The diesel electric locomotive train is the best example of this configuration. If a car has a large, externally rechargeable battery, as well as an electric motor, an engine, and an electric generator, we call the vehicle a *Plug-in Hybrid Electric Vehicle* (*PHEV*). The most common example of the PHEV as of this writing is a Toyota Prius that has been retrofitted with an aftermarket external charging regulator system and a larger battery. Toyota has already announced a new-generation PHEV Prius that will soon be available directly from the manufacturer.

6.7 PARALLEL CONFIGURATION HYBRID: TOYOTA PRIUS

The configuration is parallel because the electric motor and the engine can simultaneously propel the car. In the block diagram of figure 6.5a the two

(a) Parallel Configuration Hybrid Vehicle (with PHEV external charging)

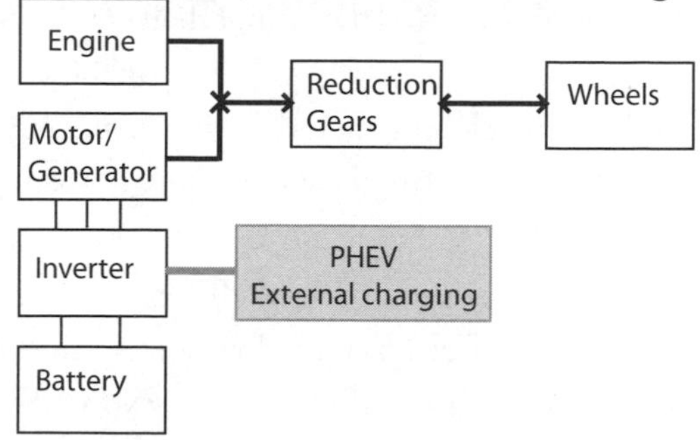

(b) Electric Vehicle

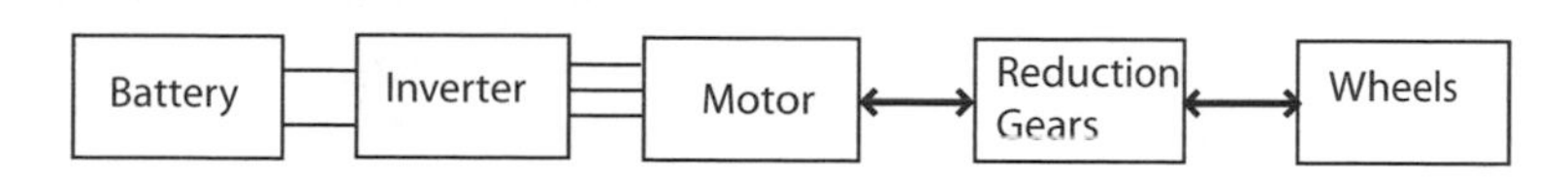

(c) Series Configuration Hybrid Vehicle

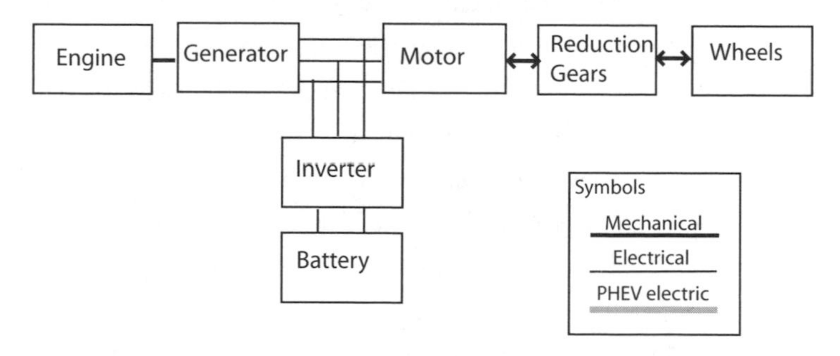

Figure 6.5 Hybrid and electric vehicle configurations. Block diagrams include the assumption that we use a three-phase AC motor because of its superior performance. An inverter performs the AC-to-DC and DC-to-AC conversions between the battery and the motor/generator.

propulsion systems are drawn in parallel. This nomenclature can get a little confusing. For example, consider the General Motors hybrid pickup truck. The GM electric motor is wound around the output shaft of the engine, before the reduction gears of the transmission. Both sources drive the same shaft. It is the fact that the engine and the electric motor can independently or simultaneously drive the wheels that makes this a parallel configuration.

The Toyota Prius is a parallel configuration hybrid. It has captured the attention of the American public as the model for all cars good and green, but is it a technology that can benefit from racing? How does it achieve its impressive mileage performance of a combined 46 mpg? In November 2004 Oak Ridge National Laboratory published a report (Evaluation of the 2004 Toyota Prius Hybrid Electric Drive System [ORNL/TM 2004/247]) that contained a summary of the techniques employed. Let's consider each technique and the implications for racing.

1. Whenever the car stops, the engine shuts down. We could employ this in a racing series where pit stops and standing starts are used. Most amateur automotive racing does not employ pit stops, whereas most professional racing does. Most road racing and oval racing in the United States employ a rolling start, but Formula 1 and its feeder series use a standing start. A standing start is dramatic, and a hybrid electric boost could add an interesting element. If we reformulate racing rules to limit the gear ratios at low speed, a conventional car would lose acceleration. On the other hand, a high-torque electric motor in the hybrid could compensate for the low ratio and give it a distinct advantage. It is possible to develop this technique for racing, but since such a small fraction of the race is spent at rest, it is not likely to make a significant change in race fuel economy.
2. The Prius employs regenerative braking converting kinetic energy into stored electrical energy. This is already in use in Formula 1, but not without some growing pains. System failures still plague the cars, including a few fires and a couple of mechanic electrocutions. Thankfully, no serious injuries have resulted. The proprietary batteries used in these systems have a one-race lifetime and then become a hazardous waste disposal problem. It also turns out that a loss of car performance due to the additional bat-

> tery and generator weight negates the gain in power that they provide. The limitations imposed by the rules have rendered it ineffective. In response, the teams have lost interest in its development. These initial mistakes by the race organizers have hurt the development of the technology, and they know it. As a result, early in the spring of 2009, the Formula 1 governing body announced plans to increase the allowed power and stored energy in the kinetic energy recovery system for 2010. If Formula 1 is interested in developing a green technology, why not make the added power and energy unlimited? Instead, they could have established minimum system weight, power, energy, and lifetimes. The teams would have gone all out seeking an advantage. In the end we would have ended up with a technology that might actually benefit passenger cars.

Techniques 3, 4, 5, and 6 below are based on the Prius's ability to decouple engine speed from car speed. For racing applications, it could give the combustion engine the ability to run at peak torque for acceleration, or at peak power for high speed and battery charging. When conditions permit, the engine also has the ability to operate at peak efficiency for improved gas mileage.

3. The Prius combustion engine is able to operate at the most efficient rpm a majority of the time. This will not be the case in race applications.
4. The electric motor provides supplemental power during acceleration under conditions where the efficiency of the combustion engine would normally be low. For example, the car accelerates from a stop using only electric power.
5. The Prius can draw power from the battery or charge the battery depending on the status of engine efficiency. A race car spends a large fraction of its time at wide-open throttle where the Prius would normally give an electric boost to acceleration. We need to keep in mind that the battery's stored energy is not limitless. At some point, the engine's resources will need to charge the battery. When does the engine have surplus power? Under braking and during cornering the engine of a conventional car is along for the ride. A hybrid could use regenerative braking and at the same time charge the batteries with the engine.

6. The Prius engine can drive the wheels directly, or it can route power through a generator to an electric motor, which can also drive the wheels.
7. The availability of supplemental battery power allows the Prius combustion engine to employ a number of efficiency techniques, including the use of a high expansion ratio thermodynamic cycle, a variation of the Atkinson cycle. Efficiency techniques include the following:

   - In the Atkinson cycle, the intake valve is held open beyond the bottom of the piston's motion during the intake stroke. This improves the efficiency of the compression stroke but produces a reduction of power at high rpm. In the Miller cycle employed by the Prius, they recoup some of the lost power in the Atkinson cycle by supercharging the inlet air. A supercharger is a compressor typically driven by a belt attached to the engine's crankshaft.
   - The engine reduces piston friction during the combustion stroke by offsetting the crankshaft to one side. This keeps the piston connecting rod more vertical during the power stroke and reduces sideways thrust of the piston against the cylinder wall.
   - They also limit the engine to a relatively low 4500 rpm, avoiding high-rpm friction losses. The power lost to friction is equal to the friction force times the piston speed. The higher the engine rpm, the higher the piston speed.

8. The Prius employs the aerodynamic Kamm/Koenig-Fachsenfeld roofline and chopped tail, sometimes called the Kamm back. With a C_d of 0.26, it possesses one of the lowest drag coefficients on the passenger car market.

You may be wondering, how can the combustion engine and the electric motor simultaneously drive the wheels and still have independent speeds? The clever solution employed by the Prius was developed by an American company, TRW Inc., back in the late 1960s and early 1970s (see U.S. patent 3566717, for example). Figure 6.6 shows a block diagram of the Prius power train.

The power train consists of an engine, two AC electric motor/generators, a battery, an inverter/converter, a set of reduction gears, and a power split device. The power split device is the planetary gear set shown in figure 6.7.

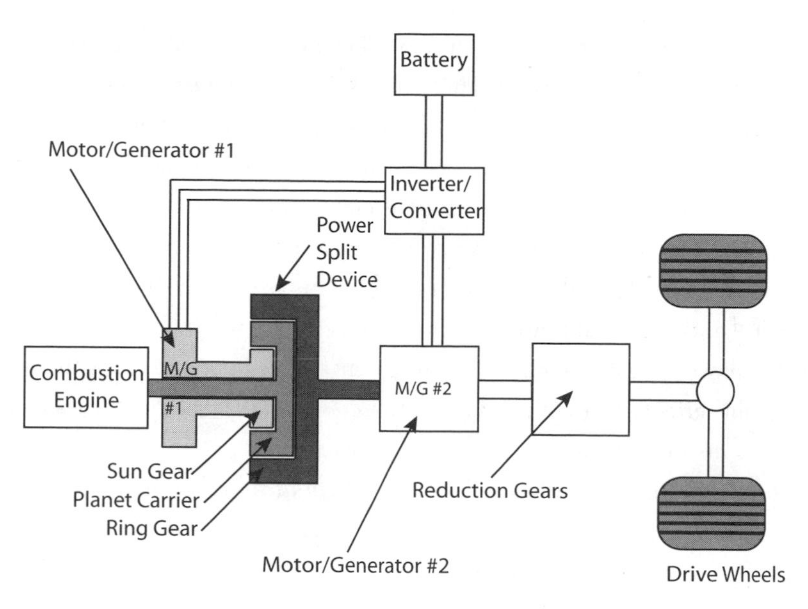

Figure 6.6 Block diagram of the Prius power train.

Planetary gears have been around for a long time. However, the method in which they are employed is unique. At the center, the sun gear is attached to Motor/Generator #1 (M/G #1). M/G #1 is primarily a generator that can charge the battery or electrically drive M/G #2. M/G#1 can also act as a starter motor for the engine. The engine drives the planetary gear assembly. As with a real planet, a planet gear can rotate about its own axis as well as around the sun gear. The combustion engine makes the planet gears orbit the sun. They connect the outermost ring gear to M/G #2 via a shaft. The shaft delivers power to reduction

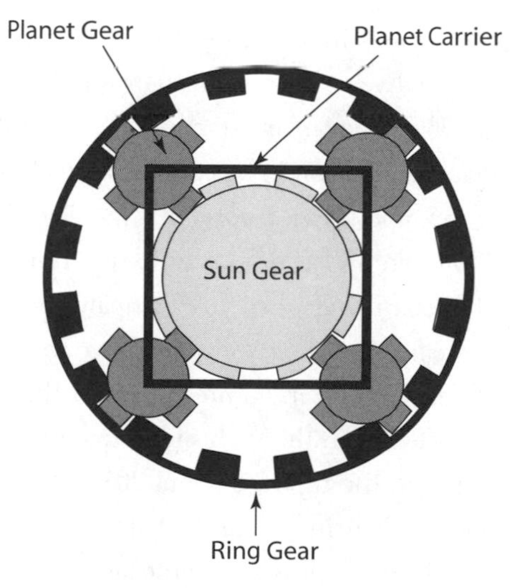

Figure 6.7 Planetary gear set.

gears using a chain. The reduction gears slow the rotation rate and deliver the power ultimately to the wheels. Thus, both M/G #2 and the ring gear can send power to the reduction gear. M/G#2 can also slow the reduction gears by acting as an electrical generator in a regenerative braking mode. If we turn the combustion engine off, the planetary assembly stops orbiting the sun. However, each planet is still free to spin on its own axis. This allows both motor/generators to rotate while the combustion engine is stopped. The flow of mechanical and electrical power is complex. It is microprocessor controlled and, for the most part, its operation is transparent to the driver. Perhaps the most noticeable effects are the silent takeoff from rest and the fact that the transmission is continuously variable.

It appears that hybrid technology could benefit from racing development. Establishing rules that will encourage manufacturers and racers is critical.

6.8 ALL-ELECTRIC VEHICLES

An all-electric race car may not be too far over the horizon. A number of technical challenges still stand in the way. Large and powerful electric motors and batteries are both heavy and bulky. For example, consider the Tesla Roadster, the first production all-electric sports car. The Tesla is powered by 6831 lithium-ion laptop computer batteries. According to the manufacturer, the battery pack weighs in at 992 lbs. Its high-power electric motor weighs an additional 115 lbs. Even so, its performance potential is intriguing. With a peak torque of 276 ft-lbs available starting at 0 rpm (fig. 6.8), its acceleration is blistering, reaching 60 mph in less than 4 s. At 14,000 rpm the car reaches its top speed of 125 mph. Tesla reports that the electric motor makes 248 hp from 5000 to 8000 rpm and 276 ft-lbs of torque from 0 to 4500 rpm. This all becomes a little confusing when you compare the quoted statistics with the dynamometer curves on their Web site. Since we already know how to analyze this type of data, let's try it. Figure 6.8 shows this torque as a function of rpm, crudely extracted from the Web site, using a ruler and a calculator. The peak torque shown in the figure is about 205 ft-lbs. Perhaps they measured the torque at the wheels using a chassis dyno. If so, the measured torque corresponds to a 25% drivetrain loss. Given the fact that the Tesla has one transmission gear, no clutch, and almost no driveline to rotate, this loss is almost unbelievably large.

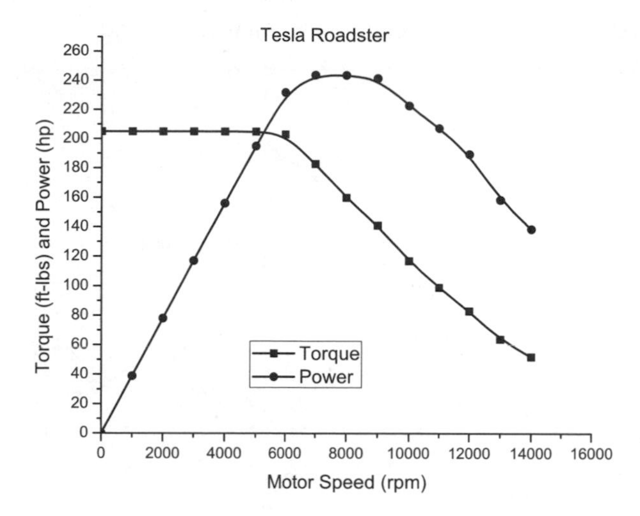

Figure 6.8 Torque as a function of electric motor rpm from the Tesla Motors Web site (www.teslamotors.com/performance/acceleration_and_torque.php). The plotted power is calculated from the torque curve.

The power curve shown in figure 6.8 is calculated using equation (2.2) and the torque curve. It comes close to the 248 hp reported by Tesla Motors.

We can test the validity of the reported torque curve by applying the techniques of chapters 1 and 2. *Road & Track* tested the 2009 Tesla Roadster and reported a curb weight of 2750 lbs. Tesla Motors reports the final reduction ratio as 8.27:1 and the rear tire size as 225/45 R17. This means that the tire radius is about 1.04 ft. Plugging this all into equations of chapter 1, we get the estimated speed as a function of time curve in Figure 6.9.

Neglecting aerodynamic drag, figure 6.9 yields a 0 to 60 mph time of approximately 4.5 s. While this is clearly in the ballpark, it does fall short. Once we have a spreadsheet generated, we can alter the input torque and estimate the required value to break the 4 s barrier. About 240 ft-lbs of torque as read from a chassis dyno from 0 to 5000 rpm is required. This corresponds to about a 13% driveline loss, down from 276 ft-lbs, a significantly more reasonable value.

This is not meant to imply that Tesla Motors is intentionally misleading us. It is probable that they left out some critical piece of information or that the webmaster is having trouble keeping up with the technical changes in

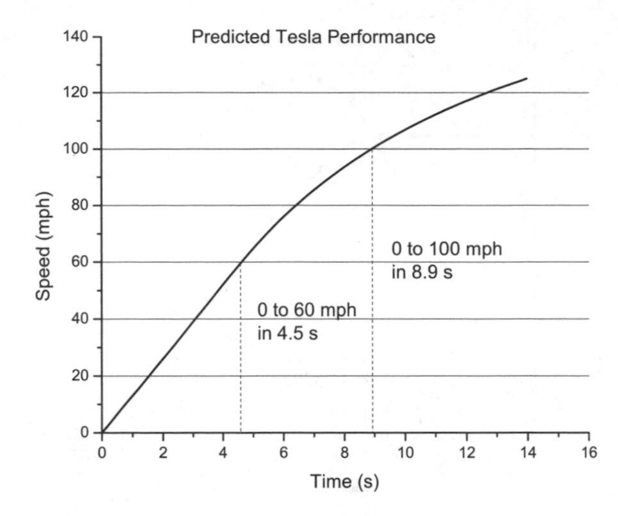

Figure 6.9 **Calculated speed as a function of time for the Tesla Roadster using torque curves from the Tesla Motors Web site.**

the car design. The important point is that the physics hasn't changed. The techniques of earlier chapters still work, and motivated readers can find these types of technical inconsistencies for all kinds of products. Now that the push is on to change our automotive paradigm to newer and greener technologies, it is more important than ever to question the data and conclusions presented to us.

Returning to the question of racing, will the Tesla Roadster make a capable road racer? The Roadster is based on the Lotus Elise, one of the lightest and most nimble sports cars on the market. The electric powertrain of the Roadster is substantially heavier and larger than the Elise. To accommodate this and to minimize the weight, the Elise was stretched a couple of inches and many components were replaced with lightweight aluminum and carbon fiber. Table 6.2 compares the statistics of the two cars. In the end, the Roadster is almost 800 lbs heavier, but substantially more powerful. The Tesla's 0 to 60 mph time of 4.0 s is awesome compared to the merely outstanding 4.7 s of the Elise. This trend in performance is backed up by the quarter-mile times. The Tesla holds that advantage up to 125 mph, where the electric motor maxes out at 14,000 rpm. The Elise continues to climb, coming close to the 150 mph mark.

TABLE 6.2
Approximate Performance Data

	Lotus Elise 2005	Tesla Roadster 2009	Notes
Propulsion	4 Cylinder Internal Combustion Vehicle	6831 Li-ion Batteries Electric Vehicle	
Approximate cost	$40,000	$109,000	
0 to 60 mph	4.7 s	4.0 s	
Top speed	150 mph	125 mph	
Skid pad	0.98 g	0.92 g	200 ft diameter circle
Curb weight	1980 lbs	2750 lbs	
Slalom	71.1 mph	68.6 mph	700 ft with 100 ft spacing
Quarter mile	13.4 s	12.7 s	
60 to 0 mph	105 ft	119 ft	
80 to 0 mph	188 ft	210 ft	
Range	224 miles	132/244 miles	Performance/Highway
Miles per gallon	26 city/38 highway	135 combined	Tesla—EPA calculated
Peak power	190 hp at 7800 rpm	248 hp 5500 to 6800 rpm	Mfr data
Peak torque	138 ft-lbs at 6800 rpm	276 ft-lbs 0 to 4500 rpm	Mfr data

Source: Data from *Road & Track*, February 2009 and August 2004.

Inertia is a beast when you are fighting to overcome it. The extra 800 lbs in the Tesla takes its toll when we consider performance items that are not controlled by the electric power plant. The lightweight Lotus corners at 0.98 g's on a 200 ft diameter circle, while the Tesla corners at 0.92 g's. Both of these centripetal accelerations are respectable, but why the difference? Is it the extra mass? Let's set up Newton's second law for uniform circular motion. It is the friction between the tire and the road that produces the centripetal acceleration. From equation (3.4) we have

$$\sum F = ma_c = \frac{mV^2}{r},$$

where the sum of the forces is equal to the mass, m, times the square of the tangential velocity, V, divided by the radius, r, of the circular path. Friction is equal to the coefficient of friction, μ, times the normal force. On a flat road with no significant aerodynamic down-force, the normal force is equal to mg, giving us

$$\mu mg = m\frac{V^2}{r} = ma_c$$

or

$$V = \sqrt{\mu gr}$$

and

$$a_c = \mu g.$$

This appears to imply that the maximum speed and centripetal acceleration are independent of mass. We need to incorporate what we learned in chapter 4. As we go around a corner, vertical load and thus normal force transfer away from the inside tire and increase the load on the outside tire. By itself, this would not be a problem, but we saw in figure 4.21 that the coefficient of friction begins to decrease when the vertical load becomes large. The overall effect is that the lateral force capability lost by the inside tires is not fully replaced by the gains in the outside tires. Thus, as the mass and the inertia of the car increase, the cornering ability decreases. The Tesla is almost 40% heavier than the Lotus, and it is remarkable that it corners as well as it does. If we threw an 800 lb block of concrete in the passenger seat of the Elise, it clearly would not turn as well as the Tesla. Part of the reason is the ability to rearrange the battery configuration and therefore lower the center of gravity. The lower the center of gravity, the less the load transfer when cornering (chap. 4) and the greater the overall cornering force.

The Tesla all-electric Roadster is an exciting demonstration of the future of green racing. It is fast as well as nimble. It is a true sports car. It has a 244 mile range based on its combined cycle EPA rating, which drops to around 132 miles when driven hard, just as we see in combustion engine cars. This makes the Tesla great for amateur sprint races, but the 3.5 hour recharge time (70 amps at 220 VAC) rules out most professional races. Charged from the electric grid where efficiencies are significantly better than that of an engine, the EPA rates the Tesla at an equivalent gasoline mileage of 135 mpg. Little or no data exist in the public domain on the Tesla battery lifetime when it is repeatedly exposed to hard driving. The company currently quotes a $12,000 battery replacement charge for the Roadster, so battery lifetime is a nontrivial question. These technical issues are the kind that race teams love. How can we optimize output, durability, and strategy to get the most from an electric car? If racers are linked with a motivated car and battery manufacturer, the possibilities are endless. Finally, we should remind ourselves that electric vehicles are not pollution free. We export the vehicle operation pollution to the location of the electric power plant that charges the car's battery. The energy consumption and pollution from production and recycling of batteries will not be small, especially in light of the fact that the battery will probably require replacement at least once in the car's life.

6.9 FUEL CELLS

A fuel cell is very similar to a battery. It depends on a chemical reaction to produce a voltage. When a battery exhausts the available chemicals, the available electrical potential or current drops and the battery is dead. A fuel cell continuously replaces the chemicals and maintains the working potential. Fuel cells have been around for a long time. Conceived and demonstrated in the 1830s, it wasn't until the 1950s that applications became practical. The most common and least polluting version uses hydrogen and oxygen. In the 1960s the Gemini and Apollo spacecrafts employed this type of fuel cell. Figure 6.10 is a sketch of a single-cell, proton exchange membrane (PEM) fuel cell. Hydrogen gas is the fuel, which releases energy when combined with oxygen. This is an electrochemical process and not combustion, which makes it more efficient.

Hydrogen gas enters one side of the cell and oxygen enters on the other. Both gases are exposed to a gas permeable conducting electrode that contains platinum or similar material, which acts as a chemical catalyst. On the hydrogen side, it separates the electrons from the single proton nucleus of the hydrogen. The electrons are conducted away by the electrode and head off to do work in an external electric circuit. Meanwhile, the proton enters the membrane. To conduct the proton, the PEM must be wet. The proton hitches a ride on the water to the opposite electrode. At the same time, electrons return from

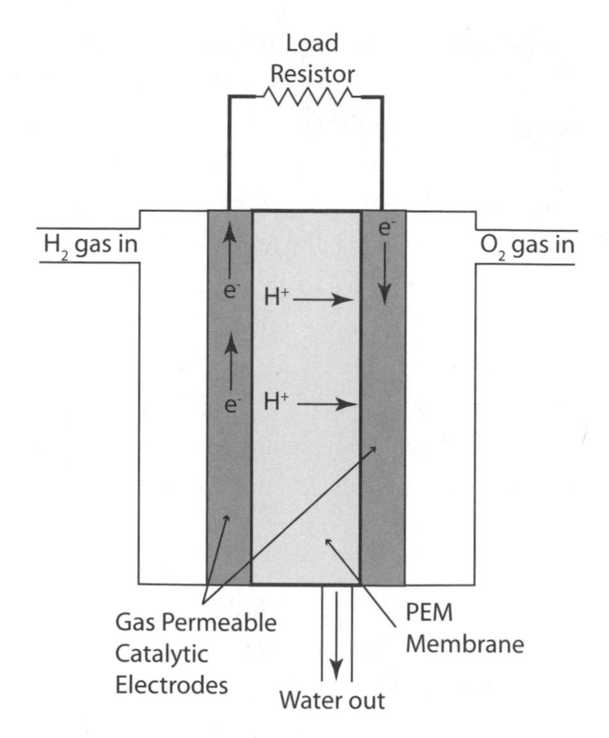

Figure 6.10 PEM fuel cell sketch.

the external circuit and, via the catalyst, combine with the oxygen, forming O^{2-}. The negative oxygen combines with the positive proton and makes water. The net chemical reaction is

$$2H_2 + O_2 \rightarrow 2H_2O.$$

This is good news. In fact, it is great news! Waste heat and water are the by-products—no greenhouse gas, no hydrocarbons, just water. A typical operating cell produces around 0.5 to 1.0 volts. Just as with batteries, if you want a higher voltage, you place the cells in series. If you want more current, you place the cells in parallel. The typical cell operates at 40%–50% efficiency, significantly better than a combustion engine.

There are some problems. Let's start big. Hydrogen is the fuel, and there is not a big cache of free hydrogen sitting around. It can be stripped from natural gas or petroleum, leaving greenhouse gas and hydrocarbons in its wake. Electrolysis can separate water into hydrogen and oxygen. However, energy must be conserved. More energy will go into the electrolysis than we will get out of the hydrogen. Hydrogen will make an excellent transportation fuel, but it is not a solution to our energy crisis. If we use the electric grid as the source of power for electrolysis, typical fuel cells are roughly one-third the grid-to-motor efficiency of a battery-powered vehicle. The theoretical potential to double fuel cell efficiency still exists. The final major problem is that we do not have a distribution system in place for hydrogen.

The big advantage of the fuel cell over the battery is the ability to refuel quickly. The time required is similar to that for a gasoline fill-up. For professional racing, this would be critical. It also answers a big consumer problem. Most of the calculations concerning the United States electric grid being able to support a large number of electric cars assume that the cars will be charged during off-peak hours. In reality, most of the test data show that people charge their cars at all hours of the day and night. A hydrogen charging station that runs on grid electricity could be programmed to produce hydrogen during off-peak hours. In turn, a fuel cell car could be refilled whenever needed.

The tide of research has begun to move away from the PEM fuel cell because of its lower grid-to-wheel efficiency. I think it is too early to make that determination. Recall from the GREET model that it is the full life cycle that matters. It is the energy, petroleum, greenhouse gas, and pollution cost over the entire life that matters. Large-scale lithium-ion batteries have just begun to be available, and their true performance potential has yet to be determined. Fuel cells are still in the running.

6.10 ALTERNATIVE FUELS

Changing fuels for our combustion engines gives us the ability to make a huge impact on pollution and greenhouse gases. The following is a brief survey of some common materials in the world of alternative fuels.

6.10.1 Diesel

Diesel fuel contains a different distribution of hydrogen and carbon molecules than does gasoline. It is less volatile and more difficult to ignite than gasoline. This affects the choice of compression ratio for our engine. Compression ratio is the maximum cylinder volume divided by the minimum cylinder volume.

TABLE 6.3
Approximate Energy Densities

Storage Media	Energy/Volume (MJ/L)	Energy/kilogram (MJ/kg)
Petroleum		
Diesel	37	46
Biodiesel	33	42
E85 ethanol	26	33
Natural gas 3600 psi	9	54
Gasoline	34	46
Hydrogen		
Liquid hydrogen	10	143
H_2 gas 10,000 psi	6	143
H_2 gas STP	11×10^{-3}	143

Note: These energy densities refer to the energy liberated when the material is combined with oxygen, either by combustion or, in some cases, by electrochemical processes. The exact values for petroleum products are reference dependent. This probably reflects differences in the composition of the petroleum product.

Because of the fuel's stability, a diesel can operate at a typical compression ratio of 20 to 1, whereas gasoline typically runs at 10 to 1 or below. Any higher than this and gasoline ignites from the heat of compression during the compression stroke. This "pre-ignition," also called knocking or pinging, is very bad for the engine. By limiting the gasoline compression ratio, the spark plug can control ignition. Diesel, on the other hand, depends on the heat of compression to ignite the fuel. The mechanical stress due to this type of ignition means that diesel engines tend to be stronger and heavier. As you can tell from table 6.3, diesel fuel has about 9% more energy per volume than gasoline.

6.10.2 Diesel Thermodynamics

We have avoided thermodynamics so far. However, in order to understand the benefits of diesel fuel, we will have to learn a little. We will limit this to a quick look at the thermodynamic cycles for gasoline and diesel fuel. We will consider them in terms of a plot of pressure as a function of volume. A cycle is a closed path on the pressure versus volume plot. To understand why this is so useful, we will consider a piston in a cylinder. Figure 6.11 is a gas-filled cylinder at pressure P. The cylinder walls form the top and the sides of the chamber, and a moveable piston of cross-sectional area A constrains the gas on the bottom.

The initial volume of the gas is the height of the gas chamber, y, times the area, A. The pressure, which is a force, F, per area, pushes on the piston surface, and the piston moves a distance Δy. Let's look at the product of the pressure times the change in volume, $\Delta V = A\Delta y$, with these definitions plugged into the equation:

$$P\Delta V = \left(\frac{F}{A}\right)(A\Delta y) = F\Delta y.$$

We already know that $F\Delta y$ is the work done by the force through a small displacement, Δy. In a plot of pressure as a function of volume, the area under the curve is the work done by the gas or on the gas.

We will consider four-stroke engines where the piston travels four lengths of the cylinder. The first stroke is a downward motion drawing air into the cylinder. We add fuel in proportion to the amount of air. Depending on the type of engine, the fuel is added during different points in the cycle. In most gasoline engines, it is drawn in with the air, but it can also be direct-injected into the cylinder. Next, we compress the air with an upward motion, called the compression stroke. Combustion of the air-fuel mixture heats the gas and raises the pressure, forcing the piston downward in the power stroke. Finally, the piston moves upward in the exhaust stroke, expelling the combustion products and any residual thermal energy. The two closed paths in figure 6.12 represent the two ideal thermodynamic cycles.

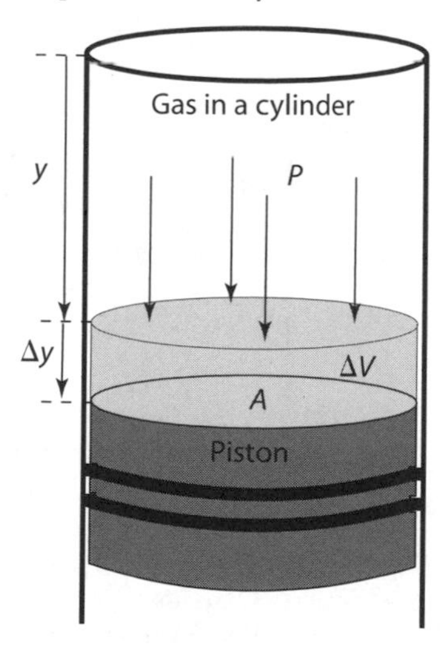

Figure 6.11 Pressurized gas in a cylinder pushes a piston and does work.

Figure 6.12 is a composite sketch of the ideal thermodynamic cycles for gasoline, called the Otto cycle, and for diesel fuel, called the Diesel cycle. Both cycles start at position 1, where the piston is at the bottom of its stroke, the volume is a maximum, and the pressure is a minimum. The next step is a compression at constant entropy and without loss of thermal energy. We will limit our discussion of entropy to saying that it is a measure of disorder of the gas within the cylinder. We call constant-entropy processes "isentropic." We call a process where no thermal energy leaks in or out of the system "adiabatic." The piston is doing work on the air during this part of the cycle. Let's finish the Otto cycle part of figure 6.12. The ratio of the volume at point 1, V_1, to the volume at point 2_O, V_{2o}, is called the compression ratio, *r*. At point 2_O, the gasoline engine is at maximum compression and the spark plug ignites the fuel, adding thermal energy at a constant volume (keep in mind that this is an idealized representation). In a gasoline engine, the throttle valve limits the amount of energy added by limiting the air. The fuel is added in proportion to the amount of air. The thermal energy raises the pressure until it is at a maximum at point 3. This pressure then acts on the piston, creating an expansion and doing work on the piston. The assumption is that the expansion is adiabatic and isentropic, again meaning that no heat leaks out and that there is no change in the disorder of the gas. From point 4 to point 1, the residual thermal energy is lost to the environment and we return to our starting point and complete the cycle. The net work done by the system is proportional to the area inside of our closed path.

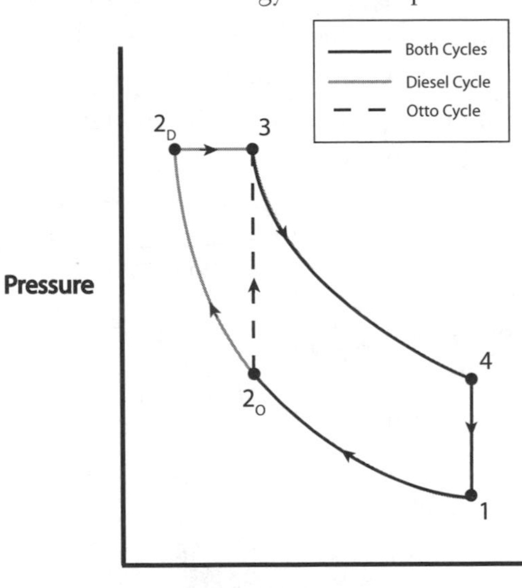

Figure 6.12 The thermodynamic cycle for ideal Otto and Diesel cycles. The black line is the same for both cycles. The gray line represents the completion of the Diesel cycle, and the dashed line is the completion of the Otto cycle.

As a student, I was confused by the ideal Otto cycle

because I could not find the exhaust and intake strokes. That is because the ideal model ignores the fact that we have to replace the air. It assumes that the air is reusable. A sketch for a real cycle would have a loop between points 4 and 1 that goes from point 4 leftward to the minimum volume and back to point 1. This would represent the exhaust and intake. Like all physics, we start with an idealized case. Usually it is much easier to do calculations on the ideal case, and there is always something to learn.

The thermodynamic efficiency, η_{TH}, is the ratio of the net work done on or by the piston to the energy added to the cylinder by the fuel and air:

$$\eta_{TH} = \frac{\text{Net Work Done}}{\text{Thermal Energy Supplied}}.$$

If all of the energy were converted to work, the efficiency would be 1.0 or 100%. This is never the case. The thermodynamic efficiency of the ideal Otto cycle can be reduced to a relatively simple expression:

$$\eta_{TH} = 1 - r^{1-k}.$$

The compression ratio, r, is V_1/V_{2O}, and k is the ratio of specific heats at constant pressure, C_P, and constant volume, C_V. C_P is the energy in kilojoules required to raise 1 kg of air 1 K at constant pressure, and C_V is the same at constant volume. For air, this ratio is $k = 1.4$. For a typical gasoline engine with a compression ratio of 8.0 to 1, we have

$$\eta_{TH} = 1 - (8.0^{(1-1.4)}) = 0.565.$$

This gives us an upper bound on the efficiency of this engine. At 56.5%, this means that at least 43.5% of the energy is lost. Where does it go? Friction, pumping losses, and heat transfer, not to mention incomplete combustion, all contribute to additional losses within the engine. We used the ratio of specific heats, k, for air, when in fact it is more complex. Air mixes with gasoline for at least part of the cycle and with combustion products for the power stroke. In general, the more complex the molecules that make up the gas, the lower the value of k, down to a minimum of 1.0. An approximate cycle average of $k = 1.3$ would yield a theoretical efficiency η_{TH} of 46.4%. Real gasoline engine efficiency is closer to 20%. We should not forget that we also have losses of energy going through the car's drivetrain. The drivetrain losses that we used in previous chapters were as little as 15%. The product of 85% times the engine's 20% efficiency yields an overall 17%. Of the chemical energy available in the fuel, about 17% becomes kinetic energy of the car. Clearly, there is room for improvement.

The Diesel cycle is similar. We start at point 1 in figure 6.12 and conduct an adiabatic isentropic compression as we did in the gasoline engine. This time we will compress the air far beyond our previous example, ending at a compression ratio of approximately $r = V_{2D}/V_1 = 20.0$. This high compression ratio heats the air enough to initiate combustion at point 2. From point 2 to point 3 we inject diesel fuel into the cylinder. The fuel burns, holding the pressure constant as the piston drops, expanding the volume. At point 3 the fuel addition is cut off. The piston continues to drop in an adiabatic isentropic expansion. At point 4 the piston is at the bottom of its travel. From point 4 back to point 1 the waste heat is rejected to the environment. We can see from figure 6.12 that the area and net work of the Diesel cycle are greater than those of the Otto cycle, even though they represent the same-size cylinder. The thermal efficiency of the Diesel cycle is affected by another important parameter, the cutoff ratio, β. This is the ratio of the volume at point 3, V_3, where the fuel is cut off, to the minimum volume, V_{2D}:

$$\beta = \frac{V_3}{V_{2D}}.$$

The thermal efficiency of the Diesel cycle is more complicated than that of the Otto cycle:

$$\eta_{TH} = 1 - \left[\left(r^{1-k} \right)_A \left(\frac{\beta^k - 1}{k(\beta - 1)} \right)_B \right].$$

The term in parentheses A is just the term from the Otto cycle. The term in parentheses B depends completely on the cutoff ratio and the ratio of the specific heats, k. To add heat in the Diesel cycle, the cutoff ratio must be greater than 1. As β approaches 1, the term in parentheses B approaches 1 and the thermal efficiency becomes the expression for the Otto cycle. The Diesel cycle uses a much higher compression ratio, around 20.0 to 1. This additional compression would raise gasoline thermal efficiency to 69.8% for $\beta = 1$. Unfortunately, gasoline can't take this compression without the occurrence of damaging pre-ignition. For a diesel with a more realistic β of 1.8, the ideal *diesel thermal efficiency falls to 65.6%, which is still a full 10% greater than the ideal Otto cycle.* If we continue to raise β, the efficiency falls. It takes a cutoff ratio of about 4.2 to drop the efficiency to that of our ideal Otto cycle, where the compression ratio was 8 to 1.

Unlike the gasoline engine, the diesel engine does not use a throttle valve in the air intake. The output power is a function of the amount of fuel injected. Injecting more fuel requires a larger β, and a larger β means a lower efficiency. Everything is a trade-off. It is also the cutoff ratio that limits the maximum diesel engine rpm. The combustion process must be complete before the piston reaches the bottom of its stroke. This leads to a practical limit of 3500 to 4500 rpm.

Diesel engines produce excellent low-rpm torque that quickly falls off at high rpm. Since power is proportional to torque times rpm, diesel engines tend to be limited in power production. Advances in turbocharging and control systems have lead to both powerful and efficient diesel cars. In the last few years both Audi and Peugeot have won the 24 Hours of Le Mans with diesels. Much of the world has embraced diesel technology, and why not? Their fuel mileage is fantastic. For example, the diesel-powered Volkswagen Polo Blue Motion 2 cranks out a combined 70.6 mpg (this is a European Union rating, not the EPA).

6.10.3 Biodiesel

Derived from vegetable oil or animal fats, biodiesel in its purest form is designated as B100. It is renewable and can be blended with petroleum-based diesel. B5 at 5% biodiesel and B20 at 20% are now available and can be used in many diesel vehicles with no modification. The Sports Car Club of America (SCCA) and Volkswagen have joined forces to create the Jetta TDI Cup racing series based on midsized, moderately priced turbodiesel passenger cars. The series is using B5 biodiesel. With young developing racers on some of the best road racing circuits in America, it is both exciting to watch and supportive of new green technologies.

It is unlikely that we could produce a major fraction of our fuel requirements from this source. However, the fact that this fuel is renewable, is derivable from waste products, and does not contribute to a net gain in atmospheric greenhouse gases makes it intriguing and a credible potential contributor to our energy needs.

6.10.4 Ethanol and Bioethanol

Pure ethanol, E100, is a relatively simple carbon chain molecule, C_2H_5OH. This is the form of alcohol found in beer, wine, and hard liquors. This liquid has a slightly lower energy density than gasoline. When complete combustion takes place with ethanol and oxygen, it leaves carbon dioxide, water, and heat. It can be chemically produced from petroleum or other hydrocarbons. The production of bioethanol from plant material has captured the attention of the energy community. Just as liquor is produced from high sugar content crops, so can ethanol fuel. The big push in the United States is to produce ethanol from corn. Brazil produces the majority of its fuel in the form of bioethanol from sugarcane. Bioethanol uses sunlight to convert atmospheric carbon dioxide and water into a combustible fuel. As such it is both renewable and ecologically friendly.

It takes energy to produce ethanol from a farmed crop, but even with corn, the energy used appears to be less than the energy obtained from the fuel. The ratio of fuel energy produced to fuel energy used in production is sometimes referred to as the energy balance. The less fertilizer and tending required, the better the energy balance. For example, sugarcane is about 4 times as efficient as corn. Interestingly enough, cellulosic ethanol made from waste wood forest products requires no tending in the growth phase. It sequesters carbon out of the atmosphere just like corn and sugarcane but requires less energy to produce than ethanol. Because the harvesting and processing equipment typically runs on petroleum products, it still results in a net release of carbon dioxide, but it is less than that of crop-based ethanol.

Perhaps the most important development in bioethanol is the technology of the company Algenol Biofuels. In June 2009, Algenol announced a demonstration project in the Sonora Desert of Mexico that should go into production in

2010. Using genetically modified blue-green algae to produce ethanol directly, they expect to be able to produce 6000 gallons per acre per year, which they claim is roughly 15 times the rate of corn-based ethanol and 7 times the rate of sugarcane. The process uses specially developed plastic balloon bioreactors and does not require arable farmland or freshwater. The algae use saltwater, CO_2, and sunlight to produce the ethanol. They claim an energy balance of 8 to 1. A total of 40 full-scale bioreactors are in operation at the company's home office in Florida. Dow Chemical, a partner in the development, is placing 3100 reactors on a 24 acre site in Freeport, Texas.

Octane rating is a measure of a fuel's ability to avoid pre-ignition or knocking. You will recall that this is where the heat of compression ignites the fuel, which is great for the Diesel cycle but bad for the Otto cycle. E100's equivalent octane rating is 110, and in certain applications as high as 130. For comparison, regular gasoline's octane is 87 and premium's is 93 (or 91 in California). This means that E100 can be run at much higher compression ratios, significantly improving its thermodynamic efficiency and power output.

Ethanol is not without problems. It is not compatible with some common materials and technologies used in the auto industry. Its low vapor pressure produces poor cold start performance, and its low energy density requires more frequent fill-ups. It can be combined with gasoline to reduce these problems. E10 (10% ethanol) and below is in widespread use to oxygenate gasoline, raise the octane rating, and reduce the amount of carbon monoxide pollution produced. These low ethanol content mixes tend to have problems with phase separation, allowing absorbed water to come out of solution. E85 ethanol at 85% has a high enough content to prevent the water from coming out of solutions. So-called flex-fuel cars are designed to run on gasoline as well as on E85. The 105 equivalent octane rating of E85 means that it is meant for a much higher compression ratio than gasoline. As a result, flex-fuel cars are inherently forced to operate with compromised performance and efficiency.

Direct injection is a new strategy for gasoline engines. Up until the last few years, gasoline was added to an engine's incoming air prior to entering the cylinder. In the first three quarters of the last century, this was accomplished using a carburetor. The carburetor contains a narrow region in the air pathway called a venturi. It creates a low-pressure, high-velocity point in the airstream. The low pressure draws gasoline into the airstream. In fact, NASCAR still uses carburetors. With the dawn of the digital age, fuel injection has virtually replaced carburetors. Fuel injection replaces the low-pressure venturi with a high-pressure spray of gasoline. As fuel injection developed, designers moved the injectors closer and closer to the intake valve. The spray patterns became more complex in both pattern and duration. In the last few years, injection of the gasoline directly into the combustion chamber at very high pressure has improved power and efficiency, as well as reducing emissions. Researchers at MIT are investigating a strategy of direct injection of gasoline followed by direct ethanol injection to suppress knocking. Because of the high effective octane number of ethanol, these engines can run with a much higher compression ratio and therefore at a higher thermodynamic efficiency.

The Indy Racing League has used ethanol as their only fuel for many years. These engines are powerful and robust. The racing is exciting. The advent of bioethanol also makes the racing a little greener.

6.10.5 Hydrogen

We have considered hydrogen-based fuel cells, but not the combustion of hydrogen. Hydrogen has a much greater energy density per kilogram than any other fuel we have considered. It has a higher octane rating than gasoline and burns at a higher temperature, leading to a greater thermodynamic efficiency. When we combust hydrogen with pure oxygen, the only waste product is water.

There are also a significant number of problems. First, we already know that hydrogen is only an energy storage media and not a new energy source. Second, there is no production, storage, or distribution infrastructure in place. From table 6.3 we see that the energy density per liter for hydrogen gas is very low. Even at pressures as high as 10,000 psi, it is still roughly one-sixth the value of gasoline. Liquid hydrogen's energy density is a little less than one-third the value of gasoline and must be maintained at −250°C to stay in a liquid state. If hydrogen is combusted with air, at its highest temperature the nitrogen component of air will react with oxygen. The resulting variety of nitrous oxides or NO_x is a form of smog. Reducing the compression ratio and temperature of combustion to eliminate the NO_x leads to reduced thermal efficiency.

Perhaps the best example of a hydrogen combustion vehicle is the BMW Hydrogen 7 concept car. Based on the 760Li, it uses direct injection and variable valve timing. This full-sized sedan is capable of running on either gasoline or hydrogen. The hydrogen tank holds 17.5 lbs, a little under 2 gallons, of liquid hydrogen that provides an effective range of 125 miles. In contrast, its 16 gallons of gasoline provide a range of 300 miles. The carbon-fiber composite liquid hydrogen tank has multiple layers of vacuum and reflective insulators that block conductive, convective, and radiative heat transport from the outside world. It provides the equivalent heat insulation of 56 ft of Styrofoam. If you are worried about hydrogen fires, BMW reports that it has completed comprehensive crash testing.

Hydrogen combustion is less efficient than the electrochemical conversion of hydrogen in a fuel cell. However, it is a viable technology that might benefit from the development that racing provides.

6.10.6 Natural Gas

Natural gas is a generic name for a collection of lightweight hydrocarbon gases. According to Union Carbide, it is primarily methane with a little ethane, nitrogen, carbon dioxide, and propane thrown in, in order of decreasing percentage. It can run at a relatively high compression ratio and therefore at a relatively high efficiency. Its waste products are primarily carbon dioxide and water. Since it is a petroleum product, the carbon dioxide waste is a new contribution to greenhouse gases. Nevertheless, because of the improved efficiency, its carbon output is about 14% less than an equivalent gasoline-powered car. From a pollution perspective, it is one of the cleanest cars on the market. Currently, natural gas is relatively inexpensive and abundant.

Compressed natural gas (CNG) already powers many vehicles. Honda makes the CNG-powered Civic GX. Its 8 gallon gas tank is pressurized to 3600 psi and provides a range of around 170 miles. With a 12.5 to 1 compression ratio, it produces an equivalent combined EPA rating of 28 mpg. With

113 hp and 109 ft-lbs of torque, it is not a sports car. Its suggested retail price is currently $6800 more than an equivalent gasoline-powered Civic, and refueling stations are few and far between. CNG is not without challenges.

Once again, we have a viable fuel that might benefit from a racing challenge.

6.11 SUMMARY

We have familiarized ourselves with a number of new automotive technologies that are "green." We were able to identify a few technologies that might benefit from racing development.

Lessons learned:

- We learned that a robust evaluation of green technologies must compare the entire fuel and vehicle life cycle to access its energy and environmental impact. The Argonne National Laboratory's GREET model is a good starting place.
- The effectiveness of regenerative braking is dependent on the energy storage mechanism. Both mechanical and electrical strategies exist. Both the amount of energy to be stored and the rate at which it is stored and used are critical when planning the system. Some racing series already employ regenerative braking.
- The brakes used in Formula 1 generate thermal energy at a rate equivalent to several thousand horsepower.
- Chemical energy densities are, in general, better than those of existing electrical energy storage devices. The volume energy density of gasoline is 3 orders of magnitude better than that of current battery or capacitor systems. The mass energy density is a factor of 100 better.
- We studied the energy-saving strategies used in hybrid vehicles and, in particular, those used in the Toyota Prius. We concluded that we could apply hybrid technology in the racing environment.
- We considered all-electric vehicles and, in particular, we looked at the Tesla Motors Roadster. We concluded that electric vehicles would benefit from development in a sprint-racing environment.
- We reviewed dynamometer data for the Tesla Roadster provided by the

company and concluded that it was inconsistent with observed performance of the car. (The car outperforms the dynamometer data.)

- We learned how the hydrogen fuel cell works and concluded that, from a refueling perspective, it was a significant improvement over recharging a battery.
- The use of alternative fuels is an excellent way to extend the effectiveness of petroleum or to replace it completely.
- Diesel fuel is thermodynamically more efficient than gasoline.
- Biofuels, such as biodiesel and bioethanol, can significantly reduce greenhouse gas emissions. Promising new production technologies are in development.
- Hydrogen is an energy storage medium and not a new source of energy. No infrastructure exists to produce hydrogen from water or to make it available to drivers.
- Combustion of hydrogen in an engine is greenhouse gas free and can be made to produce minimal pollution emissions. Production of hydrogen from petroleum releases greenhouse gases.
- Natural gas is relatively abundant, low in pollution, and produces less greenhouse gas than gasoline. CNG suffers from the same distribution problems as hydrogen.
- All the alternative fuels considered could benefit from development in a racing environment.

Conclusion

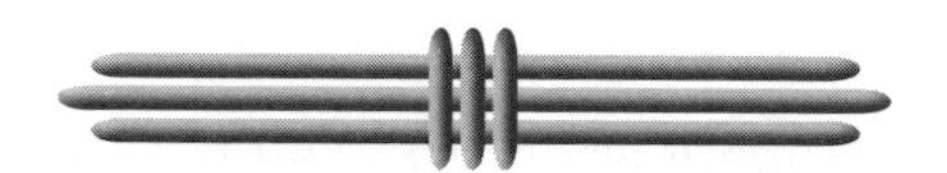

Get Off the Streets and Go Racing!

We have explored a few of the basic physics ideas behind fast driving and fast cars. Perhaps the point of this exercise, as much as any, is that first-year physics is incredibly powerful. Its insight is not limited to the world of blocks on inclined planes. Physics can be a key that unlocks our most fun activities and explains our most demanding and complex challenges.

Motorsports are truly fun, but they carry with them some degree of risk. It is foolish to bring these activities to the street. You endanger yourself as well as the people around you. Tickets and insurance can become staggeringly expensive. You can protect yourself and others, as well as your pocketbook, by taking your motorsport activities to a controlled environment.

In the world of automotive sports, the least expensive and lowest risk activity is autocross. All across the country, clubs stage these competitions in stadium parking lots and on unused runways. They are open to anyone with a driver's license, a safe car, and 30 or 40 bucks. Drivers compete against the clock. Tracks are typically less than a mile and top speeds are around 60 mph. Drivers compete by classes that match similar cars. Twisting turns and slaloms are followed by quick sweepers at the limit of traction; the emphasis is on driving skills. The courses are laid out using orange traffic cones and are evaluated by safety stewards to ensure that skidding cars have room to stop. A helmet is required, but loaners are usually available. Look for a good club, like the Sports Car Club of America (SCCA), and instructors will be available to help you with your skills. The vast majority of cars that compete are daily driven street cars. As your skills improve, you can buy a set of R-compound racing tires or move to a high class of competition with increasingly modified cars. It is exciting, challenging, and very demanding. This is entry-level racing at its finest.

If high speed is your desire and you are ready to increase the amount of risk you are willing to face, a day at the track may be your cup of tea. The SCCA calls them the Performance Driving Experience, PDX for short. Primarily intended for street-legal cars, the safety inspection criteria are slightly higher than those of autocross. The costs start at approximately $200 for a one-day event. Student drivers receive classroom and in-car instruction, typically on a full-fledged road race circuit. Passing is limited to a few zones and only when the car being passed acknowledges the pass. The driver simply points to the side of the car they expect the passer to use. Typically, they ban the timing of laps. These are driving schools, not racing schools. Successful students eventually earn the right to drive solo. Where else can you break the 100 mph barrier and not get in trouble?

Once you can solo on the track, if you absolutely have to know your times, you can begin to compete in time trials. Costs are similar to driving schools. Steps are usually taken to prevent drivers from passing or directly racing each other on track. The urge to push closer to the limit is greater once timing begins. As a result, time trials frequently require a greater degree of safety equipment. A mandatory roll bar and fire suit are common requirements. Time trials bring out everything from daily drivers to all-out track cars.

If racing door handle to door handle is your desire, club racing is your sport. This carries the highest degree of risk and the most mandatory safety equipment. Races are divided into classes to ensure close competition. A used low-power racer with an up-to-date safety inspection starts in the $5000 to

$6000 range. Costs quickly skyrocket, depending on what you want to drive. If you are willing to do your own work, you can hold cost to a minimum. The opportunity is there, and it is cheaper than you think.

Drag racing, rally racing, go-karts, drifting, motocross, or sport bikes—they all have programs designed to help you get started. Keep it off the street and do it right. While you're at it, think about the physics. It just may help you.

Suggested Reading

These are the books that I used while developing my course, The Physics of Motorsports. They were invaluable in clarifying my thoughts. I have tried to accurately attribute any unique ideas or data that may have come from these texts. They are listed in order of the significance in which they contributed to getting the physics, engineering, and driving correct. For example, *Race Car Vehicle Dynamics* is a true engineering textbook. It is a challenging book to read, but I am in awe of the author's accomplishment.

Milliken, William F., and Douglas Milliken, *Race Car Vehicle Dynamics*, Society of Automotive Engineers, Warrendale, PA, 1995.

Milliken, Douglas L., Edward M. Kasprzak, L. Daniel Metz, and William F. Milliken, *Race Car Vehicle Dynamics: Problems, Answers and Experiments*, Society of Automotive Engineers, Warrendale, PA, 2003.

Johnson, Alan, *Driving in Competition*, Bond, Parkhurst Publications, Newport Beach, CA, 1971.

Haney, Paul, *The Racing and High Performance Tire: Using the Tires for Grip and Balance*, Society of Automotive Engineers, Warrendale, PA, 2003.

Stone, Richard, and Jeffrey K. Ball, *Automotive Engineering Fundamentals*, Society of Automotive Engineers, Warrendale, PA, 2004.

Bentley, Ross, *Speed Secrets: Professional Race Driving Techniques*, MBI Publishing, Osceola, WI, 1998.

Bentley, Ross, *Speed Secrets 2: More Professional Race Driving Techniques*, MBI Publishing, Osceola, WI, 2003.

Watts, Henry A., *Secrets of Solo Racing: Expert Techniques for Autocrossing and Time Trials*, Loki Publishing, Sunnyvale, CA, 1989.

Van Valkenburgh, Paul, *Race Car Engineering and Mechanics*, published by the author, Seal Beach, CA, 2000.

Smith, Carroll, *Racing Chassis and Suspension Design*, Society of Automotive Engineers, Warrendale, PA, 2004.

Smith, Carroll, *Drive to Win: The Essential Guide to Race Driving*, Carroll Smith Consulting, 1996.

Smith, Carroll, *Engineer to Win: Understanding Race Car Dynamics*, MBI Publishing, St. Paul, MN, 1984.

Smith, Carroll, *Carroll Smith's Nuts, Bolts, Fasteners and Plumbing Handbook*, MBI Publishing, St. Paul, MN, 1990.

Smith, Carroll, *Prepare to Win: The Nuts and Bolts Guide to Professional Race Car Preparation*, Aero Publishers, Fallbrook, CA, 1975.

Smith, Carroll, *Tune to Win: The Art and Science of Race Car Development and Tuning*, Aero Publishers, Fallbrook, CA, 1978.

Donohue, Mark, and Paul Van Valkenburgh, *The Unfair Advantage*, 2nd edition, Bentley Publishers, Cambridge, MA, 2000.

Adams, Herb, *Chassis Engineering: Chassis Design, Building & Tuning for High Performance Handling*, Berkley Publishing Group, New York, NY, 1993.

Fowles, Grant R., *Analytical Mechanics*, 2nd edition, Holt, Rinehart and Winston, Austin, TX, 1970.

Tipler, Paul A., and Gene Mosca, *Physics for Scientists and Engineers*, 6th edition, W. H. Freeman, New York, NY, 2007.

Tremayne, David, *The Science of Formula 1 Design: Expert Analysis of the Anatomy of the Modern Grand Prix Car*, Haynes Publishing, Sparkford, U.K., 2004.

Parker, Barry, *The Isaac Newton School of Driving: Physics and Your Car*, Johns Hopkins University Press, Baltimore, MD, 2003.

Jurgen, Ronald K., *Electronic Braking, Traction, and Stability Controls*, volume 2, Society of Automotive Engineers, Warrendale, PA, 2006.

Haney, Paul, *Inside Racing: A Season with the Pacwest Cart Indy Team*, TV Motorsports, Redwood City, CA, 1998.

Dixon, John C., *The Shock Absorber Handbook*, 2nd edition, John Wiley & Sons, England, 2007.

Gillespie, Thomas D., *Fundamentals of Vehicle Dynamics*, Society of Automotive Engineers, Warrendale, PA, 1992.

Breuer, Bert, and K. H. Bill, *Brake Technology Handbook*, Society of Automotive Engineers, Warrendale, PA, 2008.

Gran, Dave, *Go Ahead—Take the Wheel: Road Racing on Your Budget*, Dragon Publishing, Newington, CT, 2006.

Puhn, Fred, *How to Make Your Car Handle*, Berkley Publishing Group, New York, NY, 1981.

Glimmerveen, John H., *Hands-On Race Car Engineer: Just How Important Is a Tenth of a Second?* Society of Automotive Engineers, Warrendale, PA, 2004.

Wright, Peter, *Formula 1 Technology*, Society of Automotive Engineers, Warrendale, PA, 2001.

Hoag, Kevin L., *Vehicular Engine Design*, Society of Automotive Engineers, Warrendale, PA, 2006.

Index